Journal of the International Society for Anthrozoology

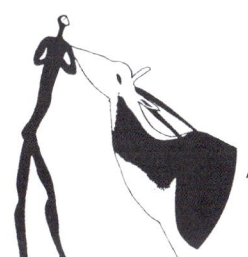

anthrozoös
A multidisciplinary journal of the interactions of people and animals

Produced in cooperation with the Humane Society of the United States (HSUS), American Society for the Prevention of Cruelty to Animals (ASPCA), WALTHAM, and the International Association of Human–Animal Interaction Organizations (IAHAIO)

Editor-in-Chief
Anthony L. Podberscek
University of Cambridge
Department of Veterinary Medicine
Madingley Road
Cambridge CB3 0ES, UK
Ph: +44-(0)1223-33 0846
Fax: +44-(0)1223-33 7610
E-mail: alp18@cam.ac.uk

Associate Editors
Penny Bernstein
Department of Biological Sciences
Kent State University Stark Campus
6000 Frank Avenue
North Canton, OH 44720, USA
Ph: +1-330-499 9600
Fax: +1-330-494 6121
E-mail: pbernstein@stark.kent.edu

Patricia K. Anderson
Dept of Sociology & Anthropology
Western Illinois University
1 University Circle
Macomb, IL 61455-1390, USA
Ph: +1-309-298 1108
Fax +1-309-298 1857
E-mail: PK-Anderson@wiu.edu

Editorial Advisory Board
Karen Allen, State University of New York at Buffalo, USA
Frank R. Ascione, University of Denver, USA
Mara M. Baun, University of Texas Health Science Center at Houston, USA
Lisa Beck, Bryn Mawr College, USA
Andrea Beetz, Friedrich-Alexander-University Erlangen-Nuremberg, Germany
Brenda Bryant, University of California Davis, USA
Rebecca Cassidy, Goldsmiths College, University of London, UK
Matthew Chin, University of Central Florida, USA
Stine B. Christiansen, University of Copenhagen, Denmark
Grahame Coleman, Monash University, Australia
Beth Daly, University of Windsor, Canada
Clifton Flynn, University of South Carolina Upstate, USA
Erika Friedmann, University of Maryland School of Nursing, USA
Kasey Grier, Winterthur Museum and Country Estate, USA
Sonya Hill, North of England Zoological Society, Chester Zoo, UK
Yuying Hsu, National Taiwan Normal University, Taiwan
Leslie Irvine, University of Colorado, USA
Sarah Knight, University of Portsmouth, UK
Garry Marvin, Roehampton University, UK
Ádám Miklósi, University of Eotvos, Hungary
Gene Myers, Western Washington University, USA
Jo-Ann Shelton, University of California, Santa Barbara, USA
Joanna Swabe, Independent Scholar, The Netherlands
Nicola Taylor, Flinders University, Australia

National Members of IAHAIO
Association Française d'Information et de Recherche sur l 'Animal de Compagnie [AFIRAC] (France)
Associazione Italiana Uso Cani da Assistenza [AIUCA] (Italy)
Belgian Association for Study & Information on the Human–Animal Relationship [ETHOLOGIA] (Belgium)
Companion Animal Research and Information Centre [CARIC] (People's Republic of China)
Delta Society (USA)
Feleos Allatbaratok Egyesulete – Association of Responsible Pet Owners (Hungary)
Forschungskreis Heimtiere in der Gesellschaft – Pets in Society Research Group (Germany)
Human–Animal Bond Association of Korea [HAB Korea] (South Korea)
Institute for Interdisciplinary Research on the Human–Pet Relationship [IEMT] (Austria)
Institute for Interdisciplinary Research on the Human–Pet Relationship [IEMT] (Switzerland)
Japanese Animal Hospital Association (Japan)
MANIMALIS (Sweden)
Petcare Information and Advisory Service (Australia)
Polish Responsible Pet Owners Association [PRPOA] (Poland)
Research in Animal Therapy and Animal Assisted Education [RETAA] (Luxembourg)
Società Italiana de Scienze Comportamentali Applicate [SISCA] (Italy)
Society for Companion Animal Studies [SCAS] (UK)
Society for the Study of Human–Animal Relations [HARS] (Japan)

Anthrozoös (ISSN 0892-7936 print; ISSN 1753-0377 online) is published four times per year by Berg Publishers, 1st Floor, Angel Court, 81 St Clements Street, Oxford OX4 1AW UK. Four issues form a volume.

2009 Subscription Rates

Print
Institutional (1 year): £152/$297; (2 years) £244/$475

Online Only
Institutional (1 year): £129/$252; (2 years): £207/$404

Free online subscription for institutional print subscribers.

Full color images are available online.

Access your electronic subscription through www.ingentaconnect.com

2009 Membership Rates

Individual: £53/$95
Society Affiliates: £53/$95
Student Affiliates: £26.50/$47.50
Corporate Membership: £159/$285
Lifetime Membership: £1,060/$1,900

Members and Affiliates of the International Society for Anthrozoology (ISAZ) receive the journal as part of their membership package.

Individual membership of the ISAZ is open to individuals currently or previously involved in conducting scholarly research within the broad field of human–animal interaction. Individuals who have an interest in the field of human–animal interactions, but who have not conducted scholarly research in the field, may apply to become Society Affiliates. For further details and to apply for membership, please see www.isaz.net.

Institutional Orders and Payment

Turpin Distribution handle the distribution of this journal. Institutional orders accompanied with payment (checks made payable to Turpin Distribution) should be sent directly to Turpin Distribution, Stratton Business Park, Pegasus Drive, Biggleswade, Bedfordshire SG18 8TQ, UK
Tel: +44 (0)1767 604951. Fax: +44 (0)1767 601640.
E-mail: custserv@turpin-distribution.com

Copyright © 2009 International Society for Anthrozoology (ISAZ), c/o Berg Publishers, 1st Floor, Angel Court, 81 St Clements Street, Oxford OX4 1AW, UK. All rights reserved. No part of the publication may be reproduced, stored in a retrieval system, or transmitted in any form or by any means, electronic, photocopying, recording or otherwise, without the prior permission of the publisher.

Indexing

Articles appearing in this journal are abstracted and indexed by Abstracts in Anthropology; Animal Behavior Abstracts; CAB Abstracts; Current Advances in Ecological and Environmental Sciences; Current Contents/Social and Behavioural Sciences; Environmental Periodicals Bibliography; Focus on Veterinary Science and Medicine; Indian Journal of Veterinary Medicine; Linguistics and Language Behaviour Abstracts; Psychological Abstracts; Referantivnyi Zhurnal: Biologiia; Science Citation Index Expanded; Social Science Citation Index; Sociological Abstracts; Veterinary Bulletin.

Berg Publishers is a member of CrossRef

Information for advertisers

Advertising orders and inquiries may be sent to:
Berg Publishers, 1st Floor, Angel Court,
81 St Clements Street, Oxford OX4 1AW UK
Tel: +44 (0)1865 241504, Fax: +44 (0)1865 791165
E-mail: enquiry@bergpublishers.com

Prepress production by Communicating Words & Images, Seattle, WA, USA.
E-mail: aptak5118@aol.com

Printed in the UK

CONTENTS

REVIEWS & RESEARCH REPORTS — 317

The Magic of Animals: English Witch Trials in the Perspective of Folklore
Boria Sax — *317*

The Emergence of "Pets as Family" and the Socio-Historical Development of Pet Funerals in Japan
Elmer Veldkamp — *333*

Reasons for Relinquishment and Return of Domestic Cats (*Felis Silvestris Catus*) to Rescue Shelters in the UK
Rachel A. Casey, Sylvia Vandenbussche, John W.S. Bradshaw and Margaret A. Roberts — *347*

Young Adults' Attachment to Pet Dogs: Findings from Open-Ended Methods
Lawrence A. Kurdek — *359*

Empathic Differences in Adults as a Function of Childhood and Adult Pet Ownership and Pet Type
Beth Daly and L. L. Morton — *371*

Social and Individual Components of Animal Contact in Preschool Children
Manuela Wedl and Kurt Kotrschal — *383*

NEWS & ANALYSIS — 397

Journal of Social Issues — Special Issue — *397*
New Books — *397*
Conferences — *398*

BOOK REVIEWS — 401

Swan — *401*

The International Handbook of Animal Abuse and Cruelty: Theory, Research and Application — *403*

INDEX TO VOLUME 22 — 407

People + Pets = Health

A simple formula. A complex relationship.

There is much to be gained in our interactions with animals, and yet so much more to be understood.

WALTHAM®, a division of Mars, Incorporated, is pleased to announce its contribution of more than $2 million to the *Eunice Kennedy Shriver* National Institute of Child Health and Human Development, a part of the National Institutes of Health, for human-animal interaction research. For more information on funding opportunities, visit **nichd.nih.gov**.

We also congratulate Rebecca Johnson, Ph.D., University of Missouri, on receiving the first annual ISAZ WALTHAM® Research Award for her study, 'Seniors fostering shelter dogs: Improving health and well-being together.'

For more information on human-animal interaction research, visit **anthrozoology.org** and **Waltham.com**.

WALTHAM® – Bringing the science to life™

The Magic of Animals: English Witch Trials in the Perspective of Folklore

Boria Sax
White Plains, New York, USA

Address for correspondence:
Boria Sax, Ph. D.,
25 Franklin Ave, Apt. 2F,
White Plains, NY
10601-3849, USA.
E-mail: Vogelgreif@aol.com

ABSTRACT This article looks at familiars in a context of folklore. Possible precedents for familiars include shamanic animal helpers, mascots of deities or saints, and local spirits, all of which are usually classified by folklorists as "grateful animals." Stories of witches and familiars are in the mainstream of European folklore, and their sole novelty lies in the fact that those figures were demonized. One result of the witch trials in England was that stories of animal sages, guides, and protectors were largely eliminated from English folklore, since these too closely resembled familiars. The persecution of those who consorted with familiars helped deprive animals of all intrinsic significance, thus opening the way for their exploitation as livestock and their humanization as pets in the centuries to come. Even today, however, familiars continue to be significant, exemplifying an intense communion between animals and human beings.

Keywords: Descartes, familiars, grateful animals, Joseph Jacobs, post-domestic society, Puss in Boots, witches, witch trials

What are Familiars?

 My wife and I were sitting one evening in New York's Central Park, when a family of raccoons crossed the path in front of us. The younger raccoons walked behind us and some began to climb up the back of our bench. Meanwhile, a large raccoon—probably the mother—stood up on two legs directly in front of us, and turned her large, luminous eyes in our direction. Just then, a small group of big, tough-looking teenage boys approached. "It's a rat!" screamed one, and they ran away. If we had been in early modern England, my wife and I might well have been charged with having a familiar. How did animals come to seem so removed from our daily lives that a group of robust teenagers could be terrified of a raccoon? Much of this separation appears to go back to the English witch trials, when the animal helpers in stories that had once been told around the fire became demonized as familiars.

The *Oxford Dictionary of English Folklore* defines "familiars" as "minor demons who, at Satan's command, become the servants of a human wizard or witch," adding that, "It is one of the distinctive features of English witchcraft that these spirits were very often thought to take the form of small animals, such as would be found around farms and homes…" (Simpson and Roud 2000, p. 118). European conjurors have always used supernatural helpers, which were often shape-shifters (Graf 1997), but the great novelty here is the form these took. Instead of terrifying monsters, familiars were creatures that had generally appeared, when noticed at all, as simply part of everyday life. Cats were the most common familiars of witches, followed in order by dogs, toads, mice, and wild birds. About a third of familiars were unidentified or fantastic animals (Serpell 2002). Familiars were, in summary, a domestication of demonic powers. They were petty officials of the infernal world, and might not even be particularly evil themselves.[1]

Continental Familiars

When the Renaissance magus Cornelius Agrippa died in 1535, rumors of witchcraft that had followed him during his lifetime became increasingly virulent. In *On the Demon-Mania of Witches (1580/2001)*, the French jurist Jean Bodin accords Agrippa a unique status among necromancers, referring to him alone as the "Master Sorcerer" and adding that he "does not deserve to be named." A few paragraphs later, however, Bodin does name him, saying:

> …the Master Sorceror instructs his disciples in every idolatry, impiety, and witchcraft. It seems though that the Academics…practiced it through ignorance and error, proceeding in good faith thinking they were doing good. But Agrippa employed it because of despicable impiety, for he was for his whole life the greatest witch of his age. And immediately after his death, writes Paulus Jovius and several others, people noticed a black dog, which he called "Monsieur" coming out of the room, which went into the Rhône, and was not seen again. (book 1.3)

On the Demon-Mania of Witches quickly became the major manual for those hunting for witches during the height of the Continental prosecutions. Agrippa would eventually be a model for the character of Faust in works by Goethe and others, with Monsieur as Faust's black poodle—a guise for the demon Mephistopheles (Ball 2006).

The philosopher René Descartes adopted a dog, to which he gave the name "Monsieur Grat." The salutation not only humanizes the animal; it also recalls, with a touch of irony, the dog Monsieur. Like Agrippa, Descartes was a skeptic, who lived as an itinerant scholar/adventurer in various European courts. Like Newton and many other figures of the seventeenth and early eighteenth centuries, Descartes straddled the worlds of hermeticism and materialistic science. For most of his life, he was widely suspected of engaging in occult studies, particularly in association with the clandestine Rosicrucian Brotherhood, and recent research confirms that this was probably the case. Always extremely cautious about the political repercussions of his works, he withheld much of his writing from publication and sometimes may even have communicated with colleagues by means of secret signs (Yates 1972; Aczel 2005).

Descartes denied that animals had reason, something he identified with the immortal soul, though not that they had the capacity for sensual perception and emotion.[2] In his *Discourse on Method*, Descartes writes, "…there is nothing which leads feeble minds more readily astray from the straight path of virtue than to imagine the soul of animals is of the same nature as our own, and that, consequently, we have nothing to fear or hope for after this life, any more than have flies and ants…" Here he is concerned about denying that human beings have souls, rather than about attributing these to animals (1641/1968, discourse 5). In a letter to the Marquis of Newcastle dated 23 November 1646, however, Descartes argues that granting a soul to one variety of animal would compel one to do the same for all, including, for example, oysters and sponges. Then, to limit the proliferation of souls, one would have to postulate metempsychosis, so that a single soul might have more than one body (Fontenay 1998).

Descartes made it far more difficult to demonize animals as companions of witches. If the souls of animals were not separate from their bodies, it would be far harder to conceive of beasts as engaging in shape-shifting and many other kinds of magic. Descartes, in other words, deprived animals of any demonic—or, for that matter, angelic—significance, thus protecting both Monsieur Grat and himself. It is in part due to the work of Descartes that familiars did not become a common feature of witch trials in at least the Protestant areas of Continental Europe (Levak 2006; Russell and Alexander 2007). Descartes' English contemporary Sir Thomas Browne disputed the contention that animals were incapable of language, among other things by observing that dogs and cats talk to witches (1646/1672).

English Familiars

In 1324, Dame Alice Kyteler of Kilkenny, Ireland became possibly the first person ever to be legally charged with having a familiar. According to the indictment, she had a demonic companion named Robin Filius Artis, who would appear to her nightly in various forms including a cat and a shaggy black dog. Alice would summon her familiar by sacrificing to it nine red cocks and nine peacock "eyes" (probably from their feathers). She was condemned to be burned alive, but escaped to England (Ledrede and Wright 1843; Russell and Alexander 2007).

The first known mention of familiars in an English prosecution was in Somerset in 1530, when the presence of a toad in the suspect's house was used as evidence against her. Familiars figured prominently in the Essex trials of 1566, 1579, and 1582. The English statute of witchcraft of 1604 made it a felony to "consult, covenant with, entertain, employ, feed or reward any evil and wicked spirit to or for any intent or purpose" (Thomas 1971, p. 443).

One year after publishing *On the Demon-Mania of Witches*, Bodin accompanied his patron the Duke of Anjou, who sought the hand of Queen Elizabeth I, in marriage. It is very likely that, in addition to his literary work, Bodin could have spread his ideas in conversations at the English court (Yates 1999; Pearl 2001). Bodin had anticipated the prominence of familiars in English witch trials by writing, "For Satan, in order to deceive men has always sought euphemisms such as 'familiar spirit,' and 'white demon,' and 'little master,' because the words 'Satan' and 'Devil' are odious" (book 2.3). The French jurist, however, thought of these household spirits as guises for Satan himself, while English witch hunters considered them be lesser, subservient demons.

Most Continental witch hunters believed that witches flew to a nocturnal Witches' Sabbath to worship the Devil. Several confessions recorded by Bodin, mostly obtained through torture, tell how the Devil would take the form of a goat, and the witches would kiss his behind. The Witches' Sabbath, however, was absent from most accounts of English witches (Gaskill 2005), partly for geographic reasons. England contains none of the remote, mountainous areas where the witches of Continental Europe are alleged to have congregated. In addition, by the height of the English witch trials, the reports of witches flying to the Devil's Sabbath were starting to seem unbelievable. The notorious English witch finder Matthew Hopkins, wrote that an investigator "utterly denies that confession of a witch, when she confesses to any improbability, impossibility, as flying in the air, riding on a broom" (1647/2004, p. 58). Familiars provided a means by which the Devil allegedly communicated with witches through intermediaries, so they did not have to encounter him at the Sabbath.

Bodin claimed that witches often have a mark of the devil on their bodies, rather like a paw print. He did not recommend using such marks as a means to identify witches, however, perhaps since he believed that these could be erased by the Devil. For the English witch-hunters Matthew Hopkins and John Sterne in the mid-seventeenth century, however, these marks became the sign of a witch. The interrogators would have the suspect's body searched for a teat, through which the familiar would allegedly suck her blood.

It is possible that the witches' mark was the result of some disease, which remains unidentified, an objection that was already made by critics at the time of the witch trials. In an attempt to distinguish the marks from hemorrhoids, Hopkins gave a description of them that was vague but still somewhat illuminating. First, he said the marks are generally in an unusual place such as the lower back. Secondly, they are calloused and insensible, for example, to pin pricks. Third, the marks are susceptible to many forms and variations (1647/2004). Calvinism, which was extremely influential in England, held that misfortunes were punishments from God, so disease could be interpreted as a sign of guilt (Gaskill 2005).

Like many practices in the English witch trials, the search for a Devil's mark seems like an inversion of Catholic practice. In 1224, when St. Francis of Assisi, the patron on animals, was fasting, a seraph appeared to him in the sky and stamped the five wounds of the Crucifixion on his hands, feet, and side. These wounds, known as the "stigmata," continued to bleed and cause pain throughout his life, and Catholics interpreted them as a bodily sign of his union with Christ (Sill 1975, pp. 186–187). The stigmata also appeared on the body of St. Catherine of Siena and on several subsequent saints. Occasionally, it might take alternative forms, for example, St. Roche had a mark in the form of a cross on his breast.

The stigmata was, like miracles and relics of saints, the sort of outward sign of holiness that Catholics prized and Protestants rebelled against. In the words of Keith Thomas, "Protestantism … presented itself as a deliberate attempt to take the magical elements out of religion, to eliminate the idea that the rituals of the Church had about them a mechanical efficiency, and to abandon the effort to endow physical objects with supernatural qualities by special formulae of consecration and exorcism. Above all, it diminished the institutional role of the Church as a dispenser of divine grace" (1971, pp. 75–76). The stigmata resembled the Devil's mark in that both were a symbol of possession. In addition, both were a continuing source of blood and suffering. The only basic difference between the two signs is they had opposite symbolic meanings, since things sacred to Catholics became diabolic to Protestants.

Other methods of detecting witches that would be used in English witch trials of the sixteenth and seventeenth centuries reflected modern scientific methods in the way they attempted to isolate and observe a phenomenon. For example, an accused sorcerer would be isolated, deprived of sleep, and then watched in the expectation that his or her familiar would appear. After being kept awake continuously for days, the alleged witch would enter a dreamlike state, during which any butterfly or mouse could appear demonic (Gaskill 2005).

Precedents for Familiars

Ever since it emerged as a field of study in the latter eighteenth and early nineteenth centuries, folklore has been what we today would call a "postmodern" discipline. Ranging across folk literature from remote times and places, folklorists have found it tantalizingly easy to identify patterns and common themes, yet these remain perpetually ambiguous and incomplete. To give just one example, almost 350 different versions of the tale "Cinderella" were already known in the nineteenth century, and they were recorded on every continent. Since then, the number of versions has far more than doubled, yet, after centuries of study, there is little scholarly consensus on the origin, pattern of diffusion, or meaning of the tale (Dundes 1982).

The striking similarity of folk literature in apparently widely separated parts of the world has long puzzled researchers. Diffusionists, epitomized by the historical–geographical school of Stith Thompson, believed that folktales had migrated around the world (1979). Psychoanalysts, particularly C. J. Jung and his followers, believed that the tales exemplified archetypal patterns that were innate in the human mind (1977). The debate between adherents of these positions continued through the much of the nineteenth and twentieth centuries.

The evolutionary relationship between genes and culture was extensively studied in the 1970s by several socio-biologists such as E. O. Wilson (Wilson 1975; Sax and Klopfer 2001).[3] Some folklorists have since tried to explain the homogeneity of folklore throughout the world through patterns of evolution that were partly cultural and partly biological (Burkert 1996; Zipes 2006). These methods work best when one considers the development of tales or motifs over several millennia, and it is very difficult to adapt them to limited historical eras.

The difficulty that we encounter in studying familiars is that they resemble so many other folkloric phenomena from shamanic animal helpers to mascots of saints, so it is difficult or impossible to trace them to any single tradition. How does one deal with enormous amounts of stimulating information which constantly elude any analytic framework? One way is to employ concepts as rich, ambiguous, and malleable as the information itself. One such concept is "shamanism," which, despite having a long history, is extremely difficult to define. Hutton has written that "Shamanism… is recognized among experts to be a scholarly construct, used to group together beliefs and activities across the world which have some apparent relationship with those observed in Siberia" (Hutton 2001, p. vii).

Most popular, and many scholarly, ideas of shamanism come from Mircea Eliade, who dominated the study in the mid to late twentieth century. Together with Joseph Campbell (1968), Eliade was one of the last scholars in the neo-Victorian tradition of such cultural anthropologists as James George Frazer (1890/1963), who ranged across eras, cultures, and disciplines to identify broad ethnographic patterns. He generalized freely, and his conclusions were initially not often challenged, simply because few scholars could even approach his prodigious erudition and industry. When he encountered material that did not fit his model, Eliade often either ignored it or treated it as an aberration (Hutton 2001).

Eliade believed that the defining characteristic of a shaman was the ability to make otherworldly journeys, especially for purposes of healing but also for other reasons including divination. He also believed that shamans characteristically use supernatural animal companions. In his words:

> The presence of a helping spirit in animal form, dialogue with it in a secret language, or the incarnation of such an animal spirit by the shaman (masks, actions, dances, etc.) is another way of showing that the shaman can forsake his human condition, is able, in a word, to "die." From the most distant times almost all animals have been conceived either as psychopomps that accompany the soul into the beyond or as the dead person's new form. (1974, p. 93)

If we regard the kingdom of Hell as the spirit world, conceived as diabolic by Christian polemicists, this conception parallels those of witches and their familiars.[4]

According to Eliade, "Shamanism in a strict sense is pre-eminently a religious phenomenon of Siberia and Central Asia" (1974, p. 4). He also stated, however, that "certain shamanic elements are found in isolation in various forms of archaic magic" (p. 6). Though expressing a preference for a narrower usage that situates shamanism in Siberia, he often refers to practices of Christians, Muslims, Buddhists, Native Americans and others as "shamanic." Although he is far from clear on this point, Eliade usually seems to believe that religious ideas are spread by diffusion rather than innate in the human mind. He does not even mention Jung or the theory of archetypes in his study of shamanism, and he frequently suggests possible channels of diffusion.

This kind of vagueness was not terribly uncommon among folklorists of the nineteenth and early to mid-twentieth century, who endeavored to grasp cultural patterns in a more holistic way than researchers generally do today. That does not necessarily make Eliade's work invalid, since it is not entirely fair to judge such writing by contemporary standards. Scholarly communities, just like any others, are subject to fashions and to cultural biases. The qualities that now make Eliade's works so prominent in popular culture are no less authentic than those that elicit increasing skepticism about them in academia. Aspects of his broadly literary approach, however, at times carry over into the work of contemporary researchers who work in a changed scholarly culture, one in which precision and consistency are paramount values.

Serpell uses the term "shamanism" loosely when he argues that "…obvious parallels suggest that at least some of the animal associations in European witchcraft had their roots in archaic shamanism, particularly the apparent belief in the witch's (or shaman's) ability to shape-shift, or to perform magical acts by sending his or her spirit out in the form of an animal" (2007, p. 625). Shape-shifting and use of animal helpers are themes found in the myths and legends of many cultures across the globe (Thompson 1955; Aftandilian and Copeland 2007), and could have come from almost anywhere. They may also have been independently invented many times, and we should not assume that any tradition featuring shape-shifting or animal helpers is necessarily archaic. To call this "shamanism" is not exactly mistaken, but it does not tell us much without some definition, or at least some clarification, of the word. To trace the development of familiars, we will need to look at more specific precedents.

One possible origin of animal familiars is the animal mascots that accompanied deities of the ancient world, which served them as helpers and sometimes as alternative forms. Zeus, for example, was accompanied by an eagle, Athena by an owl, Artemis by a deer or stag, and Apollo by a crow. Among the Norse gods, Odin was accompanied by two wolves and two ravens, while Thor was accompanied by two goats. Eliade believes these mascots were originally shamanic animal helpers (1974). Blumenberg, by contrast, believes they were the original, theriomorphic forms to which the anthropomorphic gods and goddesses occasionally reverted (1985). However that may be, these figures clearly resemble animal familiars. The pagan deities had been revived in the Renaissance, especially in occult beliefs, but also aroused suspicion of idolatry and even witchcraft as that period began to draw to a close (Yates 1999).

Bodin, a man of learning who retained at least a qualified admiration for many Greco-Roman authors of antiquity, believed that pagan deities such as Jupiter, Bacchus, or Ceres had often been disguises for the Devil. The Pythia were priestesses of Apollo at the Shrine of Delphi who would enter a trance and utter prophesies, and were consulted in crises such as wars and plagues by rulers of Athens and other Greek city-states. Bodin believed they had been witches who were possessed by the Devil (1580/2001). It would have been only a small step to regard the mascots of pagan deities as diabolic emissaries, essentially as familiars.

According to traditions that date back at least to the start of the Christian era, saints, much like pagan deities, have a special intimacy with animals (Hill 1988). St. John, for example, was accompanied by an eagle, St. Jerome by a lion, St. Anthony by a pig, St. Francis by a wolf, St. Kevin by a rook, and St. Roche by a dog, and many other saints as well were patrons of animals. In 1623, when the Cardinals were deadlocked in their effort to choose a new Pope, bees, symbol of the Barbarini family, flew in the window. The cardinals took the event as a message from God and elected Maffeo Barbarini, who soon took the name of Pope Urban VIII (Freeberg 2002). According such divine significance to animals would have been far more than enough to earn an accusation of witchcraft in England from the early to middle seventeenth century.

The first recorded mention of a familiar was an attack on the Church and, most especially, on Papal authority. In June 1303, Philip IV of France called an assembly in Paris to depose Pope Boniface VIII and accused the Pope of being a necromancer, who performed his magic by means of a familiar spirit. Previously, the relation of sorcerers with their supernatural helpers had been basically utilitarian. The demons would normally live in Hell or, at any rate, in a world apart from ours, and would only come when ritualistically summoned. The Pope's familiar spirit allegedly had an almost unprecedented sort of intimacy with its master and, though not specifically identified as having animal form, sounds like a pet. In addition to assisting in magic, the spirit allegedly lived with the Pope and followed him about (Russell 1972).

The elaborate rituals of the Catholic Church impressed Protestants not only as idolatry but also as sorcery (Briggs 1976), and the chronicle of Lancaster trials of 1612 listed "Jesuits and Papists" as witches (Potts 1613/2006, p. 146). The essential Protestant idea, which was starting to emerge at the end of the Middle Ages, was that the relationship of an individual to God should be direct—not mediated by saints, by the church, by sacraments, and certainly not by animals. The mediation with the Devil by familiars, as alleged in English witch trials, was a sort of parody of Catholic practice. This does a lot to explain why familiars played little if any role in the witch trials in Catholic regions.

For its part, the Catholic Church and the Inquisition were obsessed with heresy but only peripherally concerned with witchcraft. In 1258, Pope Alexander IV prohibited the Inquisition from any concern with charges of witchcraft, provided they did not involve heresy (Alexander IV 1258/1972). It did begin to take up cases of sorcery around the middle of the fourteenth century, as the boundary between heresy and witchcraft increasingly began to blur (Gibbons 1999). In Catholic areas, witches were held to be shape-shifters, practitioners of one of the few kinds of magic not engaged in by revered figures of the Church.

Jeffrey Russell points out that in the early modern period demons started to receive "pet names" such as "Berit" and "Robin," "Rumpelstiltzkin," "Heinnekin."[5] He believes that familiars were originally local and household spirits of indigenous European mythology, such as fairies, lars, kobolds, trolls, and leprechauns, which, despite efforts of the church, had never been incorporated into the cosmology of Christianity in any consistent way (Russell 1972, p. 53, 187; Russell and Alexander 2007). In early Saxon England, elves, who were at times capable of great mischief, had been amoral spirits associated with local features such as mountains or steams. As Christianity increasingly became the dominant religion, elves were often demonized and mentioned in exorcisms (Thomas 1971; Jolly 1996). But, also like animal familiars, the elves could be domesticated and even viewed with affection. In the Canterbury Psalter of 1147, elves, with their playful grins and pointed hats, appear mischievous but not especially evil (Jolly 1996, p. 137).

Of the historical precedents for familiars that we have looked at, all seem enticingly plausible. None of the suggested causes, however, seems entirely satisfactory by itself, at least not sufficiently so to enable us to reject the rest. Among these various sub-traditions there was perpetual contact and doubtless cross-fertilization, so perhaps it may not be terribly important which one was dominant or prior. The boundaries among these various categories are easily blurred. One might argue, for example, that St. Roche, the patron of healing, is a version of Aesculapius the Greco-Roman god of medicine, which explains why both figures are traditionally accompanied by a hound. The differences among the various kinds of animal helpers in folklore seem rather tenuous in comparison with their similarities.

Grateful Animals

Shamanic animal helpers, mascots of deities, mascots of saints, and household spirits are all manifestations of one larger folkloric tradition that may be more or less encompassed under the heading of "grateful animals." Folklorists consider this a sub-class of "supernatural helpers," which is classified by them as tale type 554 (Aarne and Thompson 1964). In a typical account of English witchcraft, the familiar asks the witch for her soul or for some other favor. The witch grants this, and the familiar then serves her in return, at times bestowing wealth but most often avenging slights. There is a covenant between the witch and the familiar, which may be either unspoken or verbally agreed. To give one typical example, from the trial of the Lancaster witches in 1612, Alizon Device was one day approached by a black dog, who offered her the power to do anything if she would only give it her soul. Alizon was intrigued and sat down, whereupon the dog jumped up and sucked at her breast, leaving an enduring mark. A short time later, a peddler refused to sell Alizon some pins, and the dog asked her what she wanted done with the man. She replied that she would like him crippled, at which point the peddler immediately fell from his horse and broke his leg (Potts 1613/2006).

Analogous bonds established with magical animals, usually by feeding them or by doing them some other favor, are found in innumerable folk and fairy tales. Take the fairy tale "Puss in Boots," which was published in different versions by Gianfresco Straparola in 1553, by Giambattista Basile in 1634, and, most famously, by Charles Perrault in 1697, spanning together approximately the period when witch trials were at their height. In the version by Perrault, a desperately poor young man plans to eat his pet cat and make a muff of its skin, but spares the animal in return for its promise to serve him. The cat first terrorizes laborers into telling the king that his penniless master is a Marquis. The cat then tricks an ogre into changing himself into a rat, eats the ogre, and gives his master the ogre's castle, enabling the young man to marry the king's daughter (Opie and Opie 1974, pp. 142–151). In the protected sphere of the wealthy in Continental Europe, the machinations of the cat appeared ironically amusing rather than diabolic, but it is easy to imagine the story retold by the prosecution at an English witch trial. One must be totally enchanted by the ambiance of the story not to notice that the cat is as dishonest and brutal as any demon.

In fact, the beginning of the confession of Elizabeth Francis, one of the Chelmsford Witches tried in 1566, sounds like a version of the same story told of a female protagonist. Elizabeth had as her familiar a cat named "Satan," which she kept in a basket and fed on bread and milk. Elizabeth told the cat that she wished to become rich, and asked specifically for sheep, which then immediately appeared in her pasture. She then asked Satan to give her a wealthy man named Andrew Byles as a husband, and the cat agreed (Williams 1566/1972).

Folklorist Max Lüthi has discussed the motif of grateful, or helpful, animals at length in reference to European folklore, in this case to Grimm's fairy tale "The White Snake," (1856/1987, tale 17):

> … first, the youth on horseback takes pity on the various animals and, afterward, feels called upon to accomplish difficult tasks and to win the hand of the princess. This is a pattern that recurs in innumerable fairy tales: the man sets forth into the unknown in search of the highest, the most beautiful, or the most valuable thing. He is set tasks which seem unsolvable, but help comes to him. In our story, the three fish he has saved get him the golden ring from the bottom of the ocean, the ants pick the millet seeds out of the grass for him, and the three ravens bring him the golden apple of life from the end of the world—the final wish of the princess. (1976, p. 68)

The difference in witch trials is that the hero becomes a villain, and the animals that help him become devils, but the general pattern of the stories is otherwise common. Accounts of witches and their familiars are fully part of the grand tradition of European and world folklore.

Burkert has argued that the motif of the grateful animals, at least in many manifestations, may be traced to archaic sacrifices, which in turn go back to the offering of meat to distract potential predators and pests from vulnerable human settlements. This has many analogues in nature such as reptiles that break off their tails to escape a pursuer or arachnids that cast off legs. A fox caught in a trap may gnaw its leg off to escape, leaving only the missing appendage behind for its assailants (1996). These instances demonstrate that sacrifice can have a pragmatic purpose, is not exclusively a product of human culture, and can attain something like the status of "instinctual" behavior. If spiders and lizards can practice sacrifice, perhaps our hominid ancestors may have done the same. The remarkable persistence of grateful animals in the tales of widely separated cultures can be explained in evolutionary terms by the fact that they reinforced a useful activity for untold millennia.

The witch trials, however, appear to have been very successful in extirpating the motif of grateful animals, which came to evoke suspicion of sorcery, from English folklore. Stith Thompson's *Motif Index of Folk Literature*, the standard reference used for tales recorded from the oral tradition, does not encompass material drawn from English witch trials. It lists grateful animals as motifs B350 though B399, but does not include a single example from England, as compared to 12 from Germany and 34 from India (1955).

The English witch trials involving familiars in the seventeenth century coincided with a repression of romances, fairy tales, and related forms of literature, which were also seen as a dangerous temptation.[6] When fairy tales were revived in England during the late eighteenth and nineteenth centuries, enthusiasts had to turn to Oriental, German, and French sources rather than English ones (Avery 2000). When Joseph Jacobs published his classic *English Fairy Tales* towards the end of the Victorian era, he anticipated that readers would not know such stories even existed. His preface begins, "Who says that English folk have no fairy tales of their own?" Jacobs goes on to say that he had "found traces" of 140 tales (1898/1967, p. v), which, however, is not much compared to what was known in many other countries. The witch trials had first appropriated the tradition of fairy tales, then, in combination with puritanical codes, largely destroyed it.

Joseph Jacobs' collection does not contain a single story with a grateful animal, while these are common in the fairy tales of Grimm (Grimm and Grimm 1856/1987; Lüthi 1976). The closest is "The Well of the World's End,"[7] which was almost certainly once a tale of a witch and her familiar (Sax 1990). In the version by Jacobs, the respective roles of familiar and witch have been reversed. Briefly, a wicked stepmother demands that a young girl bring back water in a sieve from the Well of the World's End. The girl eventually finds the well. A frog agrees to tell her how to fill the sieve, but demands in return that she do as he tells her for the entire night. Following the frog's advice, she places clay and moss in the sieve to hold the water and returns home. That night the frog comes to her home to claim his reward. She is frightened and repulsed but has to agree. At the frog's command, she chops off his head, at which he becomes a prince and marries her (Jacobs 1898/1967).

What keeps this from being a fairly conventional tale of witchcraft is not only the happy ending, but also the way in which the roles of witch and familiar are reversed. Here it is the frog, the familiar, who has done a favor for the girl, and she reciprocates by serving him. The effect of this switch is to make the girl more passive, and therefore less culpable, while the frog appears almost human from the start. Interestingly, the frog solves the problem not by magic but, much as a person would, by reasoning.

Post-Domestic Society

At this point, a word of caution seems in order. If the proliferation of writing on human–animal relations at the end of the twentieth and the start of the twenty-first centuries shows anything, it is that these relationships show enormous variety and complexity in any era.[8] Attitudes towards animals in every period of history vary greatly with class, gender, profession, and so on, but most of all they vary from one individual to the next. One must, therefore, guard against trying to sum up an age in a few sentences or making facile generalizations about historical patterns.

On the other hand, scholars must now guard against a danger that may have seemed almost inconceivable for those writing about animals only about thirty or so years ago. As anthrozoology grows more sophisticated, it becomes easier to get so caught up in recording nuances, adding qualifications, and noting special cases that any prospect of a coherent picture is lost in a mass of detail. The best chance of avoiding this, I believe, is to maintain a larger perspective. When one looks back over longer historical periods, over millennia rather than centuries, vast changes are unmistakable.

If we look at the medieval books of hours, which record everyday life in the middle ages among all social classes, it is hard to find a single picture in which animals are absent.[9] These animals are extremely varied in their roles and significance. Some such as horses, donkeys, and oxen are used for labor. Others, such as pigs, sheep, and goats are used primarily as food and for other products. Farm animals are almost wild by contemporary standards, and one may encounter them foraging around the home, in the forest, or sometimes even in the towns. One finds falcons and dogs used for hunting, as well as the game animals such as deer. There are symbolic animals from the zodiac, from fables, from heraldry, and from Christian iconography, as well as fantastic beasts that appear, so far as we can tell, to be included simply for decoration. Roosters wake people at dawn, and observation of animals such as migratory birds is used to tell the time for planting and for harvest. Animals penetrate every aspect of life and thought, from religion to industry, in highly visible ways.[10]

The books of hours have no contemporary equivalent, but newspapers provide a record of daily life. For an informal comparison, I looked for pictures of animals in the front page section of my daily paper, *The New York Times*, today (November 2, 2008). Out of roughly 175 illustrations from both news stories and advertisements, I found three pictures of animals, all in advertisements for exotic vacations, plus one additional picture showing a man dressed up as the mascot of a sports team. Obviously, this survey is not remotely scientific, but it does illustrate of how profoundly daily contact with animals has been reduced since the Middle Ages. Today we encounter a vastly smaller number and variety of animals than people did a millennium or so ago. For most people, contact is confined to a few varieties of popular pets such as dogs and cats, a handful of common birds, and the squirrels in the park. How did this state of affairs come about?

At the time of the English witch trials, England was leading a major change in human–animal relations through much of the Western world. For one thing, it was in the sixteenth and seventeenth centuries, the era when trials for witchcraft were most common in England, that pets first became a normal feature of middle-class homes, and companion animals might have provided a model for familiars. Thomas has observed that familiars, "… may have been the only friends these lonely old women (accused witches) possessed, and the names suggest an affectionate relationship." As examples he mentions the three mice who were given as familiars by Mary Hockett under interrogation by Matthew Hopkins, who were called "Littleman," "Prettyman," and "Daynty" (Thomas 1971, p. 523).

In many contexts, unsanctioned association with animals in early modern England could arouse suspicion of brutalization (i.e., the degradation of people to the level of beasts). In upper class houses, infants were not even allowed to crawl, since that suggested a four-footed beast (Klaits). Even the wearing of animal masks was considered abhorrent, and there was a constant fear of bestiality. Farmers moved their animals outside their houses into separate dwellings (Thomas 1983).

The changing role of animals in human society during the early modern period has been extensively examined by Keith Thomas in *Man and the Natural World: Changing Attitudes in England 1500–1800* (1983), one of the first books to look closely at human–animal relations not only in writings by members of social elites but also in routines of everyday life. Thomas argued that the early modern period was marked by growing popularity of pets and increasing separation of other animals from human beings. Today, however, many scholars believe that Thomas did not carry his methodology far enough. Franklin criticizes Thomas for exaggerating the homogeneity of urban culture and for failing to take sufficient account of demographic differences (1999). Such criticism is pertinent when we are attempting to give a highly nuanced picture of a particular era, but it is not as significant when we are looking at broad trends that extend over many centuries.

The early modern era, like all periods, is full of divergent trends and points of view. We should remember, however, that the witch trials, like a war, were an event that involved perhaps virtually all of English society. The victims tended to be female, poor, rural, and elderly (Levak 2006). But regardless of whether they participated as victims, accusers, attorneys, judges, clergy, police, jailers, lawmakers, chroniclers, critics, bystanders or in some other capacity, all who were even passively involved would have had some basic notions impressed upon them. These included the danger of consorting with animals in any way that suggested possible magic, both on account of the Devil and of the civil authorities.

This development contributed to a restructuring of relations between animals and human beings, which, despite the complexities of the early modern era, seems clear in the perspective of many centuries. The division between pets and farm animals was growing sharper and more intense. Livestock were being deprived of all significance beyond utility, while pets were losing all utility beyond the function of providing companionship. According to Bulliet, this polarity between pets and livestock is characteristic of "post-domestic society" where: "… people live far away, both physically and psychologically, from the animals that produce the food, fiber, and hides they depend on, and they never witness the births, sexual congress, and slaughter of these animals. Yet they maintain very close relationships with companion animals—pets—often relating to them as if they were human." It is, in other words, a society where animals are deprived of any inherent value, at least apart from that acquired through their resemblance to people.[11] Bulliet adds that post-domestic societies continue to eat meat and to use a wide variety of animal products, while shrouding the raising and slaughter of animals with a shame and secrecy that is unknown in pastoral culture (2005, p. 3).

In the view of Bulliet, the transition to post-domestic society, in which English-speaking countries led the way, involved the elimination of practices that accorded animals vestiges of mythic or religious significance.[12] This is why the English-speaking nations have been foremost in both mechanization of animal agriculture and in animal rights, institutions that can both seem foreign in some countries such as Japan (2005).[13] In Bulliet's words:

> And this in turn may foreshadow a very different future for Japanese human–animal relations than can currently be visualized for the postdomestic West. Postdomesticity as it has so far developed problematizes the human–animal divide in its science, its philosophy, and its politics. But it never truly erases it. A companion animal or a fictional flock can be humanized and animals can be seen as metaphors for human dilemmas. But for all the talk of humankind being a species of animal, the legacy of postdomesticity has shut the door on specific humans actually being or becoming animals …. The many millennia of the domestic era effectively erased once vital memories of animals and human–animal hybrids as mediators or dwellers in the unseen spirit world. (Bulliet 2005, pp. 219–220)

Industrialized agriculture views animals as commodities, while the animal rights movement humanizes them, but neither recognizes their alterity.[14]

The condemnation of witches for concourse with animal familiars comes at the beginning of post-domestic society in the early modern period. One place where the numinous associations of animals are unmistakable is in folk and fairy tales throughout the world, in which beasts not only talk like human beings but also possess mysterious powers and, in consequence, can command respect. They are, among other things, sages, supernatural guides, sacrificial offerings, oracles, and guardian spirits or, in other words, mediators between us and the realm of spirits. Their stories are possibly our most intimate, as well as the most archaic, expression of the bonds between animals and human beings (Nitschke 1977; Burkert 1996). As we have seen, one effect of the witch trials was to suppress a common folkloric motif, that of grateful animals that return to serve the hero or heroine. Without a sense of animals in that role, even much Christian symbolism such as the image of Jesus as the sacrificial lamb becomes difficult to comprehend. Animals lost nearly all mythic significance, leaving them little importance beyond companionship and utility. This change facilitated the expansion of pet-keeping and the industrialization of animal husbandry in the eighteenth through twentieth centuries.

Conclusion

What differentiates accounts at English witch trials from other stories of supernatural animal helpers is fundamentally neither the action nor the characters, which are almost commonplace, but their valuation and interpretation. It is, in other words, the demonization of protagonists and their assistants which, in most other contexts, had appeared neutral or even benign. Familiars do not differ markedly from many figures in European and world folklore, including shamanic helpers, local spirits, mascots of deities or saints, and theriomorphic guides in fairy tales. The sole uniqueness of familiars lies in the fact that both they and their human associates were demonized. The persecution of witches for their association had the effect of depriving animals of spiritual significance apart from their resemblance, usually as pets, to human beings. But why were familiars able to inspire such intense fear?

The ability of owls to spot small animals from an immense distance or of dogs to track animals in a forest seems amazing to us even today. It may have been even more astonishing before people realized that animals have sensory capacities not found in human beings. The concept behind familiars is to use animals as a means to enter different perceptual worlds, to journey into realms beyond the mundane patterns of our daily lives. This is such a simple and basic idea that it does not require a continuous tradition, but is easily rediscovered.

At the end of his study of Shamanism and Eliade, Ronald Hutton suggests that the ability to commune with alternative worlds may be something found among individuals in all societies, but that some societies encourage this while others regard it as either threatening or insane (2001). Something of the sort might be said of intense communion between animals and human beings. The ability to develop such relationships is less a historical development than an abiding part of the human condition. In almost every culture or system of belief, relationships between human beings and animals can have a kind of magic.

Acknowledgements

I would like to thank my wife, Linda Sax, who read over the manuscript and made some valuable suggestions. My gratitude also goes to Editor Anthony Podberscek and the two anonymous reviewers from *Anthrozoös*, not only for their very helpful suggestions but also their patience with what has been, admittedly, a somewhat complicated project.

Notes

1. Girard (1993) has maintained that all violence has a sacrificial dimension, and this certainly applies to the scapegoating of alleged witches. In English trials, the entire burden of guilt was almost invariably laid upon the witch, and there is rarely any mention of a familiar being punished, which means that animal familiars are denied even the possible status of being sacrificial victims.
2. Descartes, whose world view was still steeped in Neoplatonism, would have regarded all things as animated, so his comparing animals to machines (1641/1968) would not have seemed to deny their capacity for emotion and sensation. It was only when thinkers read his work in a context of a more radical materialism that this implication seemed reasonable. In replying to objections to his *Meditations on First Philosophy*, Descartes specifically states that, "… I have never denied that animals have what is commonly called life, or a bodily soul, or organic senses," and similar statements are found in his letters (Descartes 1637/2008, p. 203). It is also noteworthy that Descartes always used the word "animal," which literally means "having a soul," rather than the near synonym "brute" (both are the same in French and English), which suggests a purely physical existence (Fontenay 1998). For more on the way Descartes' position on this point has been misrepresented, see Sorell (2005).
3. In the 1920s, cultural anthropologist Walter Scheidt discovered the co-evolution of genes and culture, thus providing a means by which the diffusionist and psychoanalytic positions could be reconciled (Scheidt 1930). The school of anthropology that he founded, known as Culture Biology (Kulturbiologie) became, however, associated with the Nazi regime, and, in consequence, was suppressed and then largely forgotten after World War II. More research is needed to determine how much Sociobiology was influenced by Culture Biology, perhaps through indirect channels.
4. This comparison is very clearly implicit, if not quite stated, in the discussion by Serpell (2002).
5. Serpell lists the names given to familiars in English witch trials, and almost all of them could at least be interpreted as endearments (2002).
6. "Obviously, England after 1688 was not entirely a police state, but the laws banning certain types of amusement in the theater, literature, and the arts based a far-reaching effect of the populace. In particular, the oral folk tales were not considered good subject matter for the cultivation of young souls …" (Zipes 1987, p. xiv).

7. The collection does contain other stories in which a person is helped by an animal, notably "How Jack Went to Seek his Fortune," "Jack and the Golden Snuff Box," and "Whittington and his Cat" (Jacobs 1898/1967), but the power and autonomy of these animals is very limited. I do not include these as stories of grateful animals, since the assistance by animals is not the result of a favor or the terms of an agreement. They simply show a person using animals in ingenious ways.
8. This is most emphatically the case in early modern England, as Erica Fudge, for example, has conclusively shown in *Perceiving Animals: Humans and Beasts in Early Modern English Culture* (2002).
9. The most famous example is *Tres Riches Heures of Jean, Duke of Berry*, produced at the beginning of the fifteenth century. For a reproduction of it, see Meiss (1976).
10. For a discussion of animals in art of the Middle Ages, see Benton (1992). For a survey of the roles played by animals in medieval thought and in daily life, see Salisbury (1994).
11. An interesting illustration is the way people with companion animals are now regularly referred to as "pet parents," giving the pets nearly human status.
12. "When European colonists imported European domestic animals to lands suitable for stock raising and devoid of indigenous pastoral traditions, the imported stock multiplied to form large herds of grazing animals, many of them feral. Those who sought to exploit these herds thought only in terms of selling animal products: wool, meat, and hides" (Bulliet 2005, p. 181).
13. "… [in Japan] after a century of exposure to the West, industrial growth, and urbanization, culminating in the spread of American cultural practices introduced by the post World-War II military occupation …, by 1965 annual per capita meat consumption, excluding seafood, was still only 14 pounds, a level comparable to that of some of the world's poorest countries. Thirty years later, after an astounding economic boom had transformed Japan to a leading economic power, Japanese meat consumption had risen to 67.5 pounds, but this figure was still little more than one third of the American level." But the Japanese have—in spite or, rather, because of—their low meat consumption, little comprehension of the veganism that is promoted in England and America. (Bulliet 2005, pp. 211–212).
14. Deleuze and Guattari write of "becoming animal"(2007), but their usage of the phrase is loose and superficial. For one thing, it is based on the abdication of identity, and without identity transformation is impossible.

References

Aarne, A. and Thompson, S. 1964. *The Types of the Folktale: A Collection and Bibliography*. Helsinki: Academia Scientiarum Fennica.

Aczel, A. D. 2005. *Descartes' Secret Notebook: A True Tale of Mathematics, Mysticism, and the Quest to Understand the Universe.* New York: Broadway Books.

Aftandilian, D. and Copeland, M. W. 2007. Shapeshifting: From werewolves to Japanese Kitsune to Harry Potter. In *Enclyopedia of Human–Animal Relations, Vol. 2,* 607–611, ed. M. Bekoff. London: Greenwood Press.

Alexander IV, P. 1258/1972. Magic and the Inquisition. In *Witchcraft in Europe 1100–1700*, 79, ed. A. C. Kors and E. Peters. Philadelphia: University of Pennsylvania Press.

Avery, G. 2000. British and Irish fairy tales. In *The Oxford Companion to Fairy Tales,* 66–77, ed. J. Zipes. New York: Oxford University Press.

Ball, P. 2006. *The Devil's Doctor: Paracelsus and the World of Renaissance Magic and Science*. New York: Farrar, Straus and Giroux.

Benton, J. R. 1992. *The Medieval Menagerie: Animals in the Art of the Middle Ages*. New York: Abbeville Press.

Blumenberg, H. 1985. *Work on Myth*, trans. R. M. Wallace. Cambridge, MA: MIT Press.

Bodin, J. 1580/2001. *On the Demon-Mania of Witches*. trans. R. A. Scott. Toronto: Centre for Reformation and Renaissance Studies.

Briggs, R. 1976. *Witches and Neighbors: The Social and Cultural Context of European Witchcraft*. New York: Viking.

Browne, T. 1646/1672. *Pseudodoxia Epidemica, or, Enquiries Into Very Many Received Tenents and Commonly Presumed Truths Together with the Religio Medici*. London: J. R. for Nath. Ekins.

Bulliet, R. W. 2005. *Hunters, Herders, and Hamburgers: The Past and Future of Human–Animal Relationships*. New York: Columbia University Press.

Burkert, W. 1996. *Creation of the Sacred: Tracks of Biology in Early Religions*. Cambridge, MA: Harvard University Press.

Campbell, J. 1968. *The Hero with a Thousand Faces.* Princeton, NJ: Princeton University Press.

Deleuze, G. and Guattari, F. 2007. Becoming-Animal. In *The Animals Reader: The Essential Classic and Contemporary Writings*, 37–50, ed. L. Kalof and A. Fitzgerald. New York: Berg.

Descartes, R. 1637/2008. *Meditations on First Philosphy With Selections from the Objections and Replies.* trans. M. Moriarty. New York: Oxford University Press.

Descartes, R. 1641/1968. *Discourse on Method and The Meditations.* trans. F. E. Sutcliffe. New York: Penguin.

Dundes, A. ed. 1982. *Cinderella: A Casebook.* Madison, WI: University of Wisconsin Press.

Eliade, M. 1974. *Shamanism: Archaic Techniques of Ecstasy.* trans. W. R. Trask. Princeton: Princeton University Press.

Fontenay, E. D. 1998. *Le silence des bêtes: La philosophie à l'épreuve de l'animalité.* Paris: Fayard.

Franklin, A. 1999. *Animals and Modern Culture: A Sociology of Human–Animal Relations in Modernity.* London: Sage.

Frazer, J. G. 1890/1963. *The Golden Bough: A Study in Magic and Religion.* New York: Macmillan.

Freeberg, D. 2002. *The Eye of the Lynx: Galileo, His Friends, and the Beginnings of Modern Natural History.* Chicago: University of Chicago Press.

Fudge, E. 2002. *Perceiving Animals: Humans and Beasts in Early Modern English Culture.* Chicago: University of Chicago Press.

Gaskill, M. 2005. *Witchfinders: A Seventeenth-Century English Tragedy.* Cambridge, MA: Harvard University Press.

Gibbons, J. 1999. The Great European Witch Hunt: What if everything you thought you knew about the Burning Times turned out to be wrong? *PanGaia* (21): 25–34.

Girard, R. 1993. *Violence and the Sacred.* trans. P. Gregory. Baltimore, MD: Johns Hopkins University Press.

Graf, F. 1997. *Magic in the Ancient World.* Cambridge, MA: Harvard University Press.

Grimm, J. and Grimm, W. 1856/1987. *The Complete Fairy Tales of the Brothers Grimm.* trans. J. Zipes. New York: Bantam Books.

Hill, R. 1988. Saints, beasts, and legal order in the Middle Ages. *Anthrozoös* 1(2): 65–70.

Hopkins, M. 1647/2004. In T*he Discovery of Witches: A Study of Master Matthew Hopkins Commonly Called Witch Finder 1647*, 49–62, ed. M. Summers. Whitefish, MT: Kesssinger Publishing.

Hutton, R. 2001. *Shamans: Siberian Spirtuality and the Western Imagination.* New York: Hambledon and London.

Jacobs, J. ed. 1898/1967. *English Fairy Tales.* New York: Dover.

Jolly, K. L. 1996. *Popular Religion in Anglo-Saxon England: Elf Charms in Context.* Chapel Hill, NC: University of North Carolina Press.

Jung, C. G. 1977. Approaching the unconscious. In *Man and His Symbols*, 1–94, ed. C. G. Jung and M. L. v. Franz. New York: Dell.

Klaits, J. 1985. *Servants of Satan: The Age of the Witch Hunts.* Bloomington, IN: Indiana University Press.

Ledrede, R. D. and Wright, T. 1843. *Dame Alice Kyteler, Persecuted for Sorcery in 1324.* London: John Bowyer Nichols and Son/Camden Society.

Levak, B. P. 2006. *The Witch-Hunt in Early Modern Europe.* New York: Pearson/Longman.

Lüthi, M. 1976. *Once Upon a Time: On the Nature of Fairy Tales.* Bloomington, IN: Indiana University Press.

Meiss, M. 1976. *Tres Riches Heures of Jean, Duke of Berry.* New York: Abbeville Press.

Opie, I. and Opie, P. eds. 1974. The Master Cat or, Puss in Boots. In *The Classic Fairy Tales.* New York: Oxford University Press.

Pearl, J. L. 2001. Introduction. In *On the Demon-Mania of Witches*, 9–34, J. Bodin, trans. R. A. Scott. Toronto: Centre for Reformation and Renaissance Studies.

Potts, T. 1613/2006. *The Wonderful Discovery of Witches in the County of Lancaster.* Charleston, SC: Bibliobazaar.

Russell, J. B. 1972. *Witchcraft in the Middle Ages.* Ithaca, NY: Cornell University Press.

Russell, J. B. and Alexander, B. 2007. *A History of Witchcraft: Sorcerers, Heretics & Pagans.* New York: Thames and Hudson.

Salisbury, J. E. 1994. *The Beast Within: Animals in the Middle Ages.* New York: Routledge.

Sax, B. 1990. *The Frog King: On Legends, Fables, Fairy Tales and Anecdotes of Animals.* New York: Pace University Press.

Sax, B. and Klopfer, P. H. 2001. Jacob von Uexküll and the anticipation of sociobiology. *Semiotica* 133/134: 767–768.

Scheidt, W. 1930. *Kulturbiologie: Vorlesung für Studierende aller Wissensgebiete*. Jena: Verlag von Gostav Fischer.

Serpell, J. A. 2002. Guardian spirits or demonic pets: The concept of the witch's familiar in early modern England. 1530–1712. In *The Animal–Human Boundary*, 157–190, ed. A. N. H. Craeger and W. C. Jordan. Rochester, NY: Rochester University Press.

Serpell, J. A. 2007. Witchcraft and animals. In *Encyclopedia of Human–Animal Relations. Vol. 2,* 622–625, ed. M. Bekoff. London: Greenwood Press.

Sill, G. G. 1975. *A Handbook of Symbols in Christian Art*. New York: Collier Books.

Simpson, J. and Roud, R. 2000. *Oxford Dictionary of English Folklore*. Oxford: Oxford University Press.

Sorrell, T. 2005. *Descartes Reinvented*. New York: Cambridge University Press.

Thomas, K. 1971. *Religion and the Decline of Magic*. New York: Charles Scribner's Sons.

Thomas, K. 1983. *Man and the Natural World: Changing Attitudes in England 1500–1800*. London: Allen Lane.

Thompson, S. 1955. *Motif-Index of Folk-Literature; A Classification of Narrative Elements in Folktales, Ballads, Myths, Fables, Mediaeval Romances, Exempla, Fabliaux, Jest-Books, and Local Legends*. Bloomington, IN: Indiana University Press.

Thompson, S. 1979. *The Folktale*. New York: AMS Press.

Williams, C. 1566/1972. The confessions of the Chelmsford witches England. In *Witchcraft in Europe 1100–1700*, 229–235, ed. A. C. Kors and E. Peters. Philadelphia: University of Pennsylvania Press.

Wilson, E. O. 1975. *Sociobiology: The New Synthesis*. Cambridge, MA: Belknap Press of Harvard University Press.

Yates, F. 1972. *The Rosicrucian Enlightenment*. New York: Routledge.

Yates, F. 1999. *The Occult Philosophy in the Elizabethan Age*. New York: Routledge.

Zipes, J. 1987. Introduction. In *Victorian Fairy Tales: The Revolt of the Fairies and Elves*, xi–xxiv, ed. J. Zipes. New York: Routledge.

Zipes, J. 2006. *Why Fairy Tales Stick: The Evolution and Relevance of a Genre*. London: Routledge.

The Emergence of "Pets as Family" and the Socio-Historical Development of Pet Funerals in Japan

Elmer Veldkamp
Department of Cultural Anthropology, The University of Tokyo, Japan

Address for correspondence:
Elmer Veldkamp,
Town Heights Yoshida A101,
Gakuen-higashimachi 3-7-28,
Kodaira, Tokyo,
187-0043 Japan.
E-mail:
elmer.veldkamp@gmail.com

ABSTRACT Discussions on the popularity of pet animals in present day Japan tend to stress the position of pets in a family as resembling that of human family members. In this paper, I investigate this claim by analyzing the meanings that have been attributed to animal burial and pet graves in modern history. This paper presents data on the emergence of animals as modern pets during the early twentieth century and the way their position in the family was expressed, by looking at inscriptions on pet tombstones and at the structure of posthumous care for dogs and cats. These animals used to have a strong spiritual connotation in pre-modern Japanese society, and spiritual vengeance by animal spirits was warded off by performing memorial services for these animals, which resembled rituals for untimely human death (also considered a potential source of spiritual harm). Field data from two pet cemeteries established in the early twentieth century suggest that these rituals gradually changed to the pet funerals that are common today, but still the motivations and interpretations that underlie pet funerals and memorial services in Japan are diverse. Although pet funerals nowadays are very similar to human funeral ritual, spiritual elements in the attitude towards animals have not entirely disappeared. Rapid urbanization after World War II brought along a surge in demand for pet cemeteries, requiring pet cemeteries to adopt new strategies in the form of locker-style ossuaries to efficiently store large amounts of pet remains in a limited space. Commercialization of pet funerals in the urban environment has gone hand in hand with a tendency towards shorter mourning cycles for the owners, suggesting that the significance of animal funerals in Japan has shifted from prayer for the animal soul to a way of expressing grief by the pet owner.

Keywords: dōbutsu kuyō, Japan, memorial service for animals, pet as family, pet cemetery

Local Images of Pets: Elements of the Old and New in Pet Funerals in Japan

This paper looks at the intertwinement of traditional and modern aspects of pet funerals and memorial services in Japan. A multi-disciplinary field such as anthrozoology, which is young and under continuous development, can profit from non-Western native views and data regarding human–animal problems by looking at qualitative differences between the cultures and attitudes among regions. This paper looks at one local view of human–animal relationships and the social, historical, and cultural circumstances that have accommodated changes in attitudes towards small animals (which later became pets), by discussing modern *and* traditional meanings of funerals for dogs and cats in Japan.

According to the statistics of 2007, about 12.6 million pet dogs and 10.2 million pet cats live in Japan ("Fourteenth national survey on ratios for dog and cat keeping," Japanese Pet Food Manufacturers Association, October 2007. <http://www.jppfma.org/shiryo/shiryo-set.html>). Even though the Japanese media have made frequent mention of a "pet boom" in recent years, the statistics do not confirm this. Data from the Ministry of Health, Labour and Welfare show that the number of dogs registered per year has increased from roughly 3.2 million in 1975 to over 6.7 million in 2007, showing a gradual increase rather than explosive growth ("Change per year of the number of registered and vaccinated dogs," Ministry of Health, Labour and Welfare. <http://www.mhlw.go.jp/bunya/kenkou/kekkaku-kansenshou10/02.html> Accessed January 2009). Similarly, data from a survey by the Ministry of the Environment suggest that the percentage of pet owners has remained stable. In 1974, 41.7% of the participants were pet owners, but figures between 1979 and 2003 were from 33 to 37% ("Public opinion survey regarding animal protection," Ministry of the Environment. <http://www8.cao.go.jp/survey/h15/h15-doubutu/index.html> Accessed January 2009).

Sociologist Yamada Masahiro suggests that it is actually the consciousness and attitudes of pet keepers that have changed significantly since the 1990s. These changes are exemplified by constant concern about the pet's health, by conscious "heart-to-heart" communication with the pet, and by the intention to make the pet feel satisfied and happy. Nowadays, many of these pets share the living space with their owners 24 hours a day, becoming an indispensable part of people's life rhythms. This development has given rise to the expression that pets are increasingly regarded as "family members." The image of certain animals changed from functionality in rural society to a relationship that relies heavily on the expression of intimacy towards the pet (*kawaigaru*) (Yamada 2004).

Japanese pet owners are no exception in ascribing a parent role to themselves and treating pet animals as if they were children. Framing love and care in this way shows the reverse side of the social position of pet animals, as well: the pet is treated as "one of the family," but often it is not allowed to attain "full personhood" (*ichininmae*) within the family structure. Keeping the pet in its role as a child may even help to reinforce this relationship of intimacy; hence, the dressing up of pets, and the broad range of products and services from luxury pet foods and pet clothing to fitness parlors and health clinics that allow humans to act out this intimacy.

The joy a pet provides its owner during its lifetime appears to be proportional to the shock and drama that hits owners when their pet dies. The pet funeral business is a significant phenomenon in Japan's pet industry, with over 80 pet cemeteries in the Tokyo area alone. The first popular pet cemeteries in Japan were established at the beginning of the twentieth century, but most date from the late 1960s to the 1970s, the majority being private businesses and a smaller number being managed by, or connected to, Buddhist temples.

The socio-historical background that led to these developments was partly formed by the Westernization of Japanese views of animals. Starting with the introduction of animal protectionism during the early twentieth century, from the 1920s onward systematic training and mobilization of horses and dogs in the Imperial Army took off, and intimate relationships between humans and animals came to be displayed to the public through national propaganda. Along with these developments, particular pre-modern Japanese views of animal death may have provided fertile soil for pet cemeteries to flourish.

The next section of this paper compares and assesses the relationship between posthumous rituals for small animals in pre-modern times and for pets in modern times. I will then focus on differences in the social and cultural backgrounds that provide context for the performance of these rituals. After that, the historical development of pet graves and the narrative of pets as "family members" in the changing Japanese family (*kazoku*) of the twentieth century is illustrated through data from two early Japanese pet cemeteries. Then I will look at the increase in posthumous rituals for non-human animals during the post-war period and their relation to pet funerals. The last section discusses the influence of modern factors such as urbanization and individual animal ownership on views of animals and the socio-historical development of pet funerals during the twentieth century.

Posthumous Ritual for Dogs and Cats in Japanese Folklore

In Japanese folklore, small animals such as cats and dogs were believed to be able to travel freely between the here-and-now and the afterworld, and to possess the power to wreak spiritual vengeance (*tatari*) on people. The "Dictionary of Folkloristics" (*Minzokugaku jiten*, Minzokugaku Kenkyūjo, ed. 1952) mentions spiritual possession by a "dog deity" (*inugami*), and in the entry for "animal memorial services" (*dōbutsu kuyō*), the use of "dog grave markers" (*inusotoba*) in prayer for safe childbirth is mentioned. For rituals performed by hunters for their thousandth animal killed (*senbiki iwai*), and by fishermen for fish and whales, Yanagita says "… it is highly probable they [memorial services; note by author] exemplify a consciousness older than and separate from Buddhist thought." Posthumous care for animals was already a subject of attention in Japanese culture before Buddhist terminology and form took over (which is the case with Japanese pet cemeteries).[1]

The association between dog death and childbirth is especially present in folklore in the area around Tokyo. Dogs were considered to give birth easily, and a "dog memorial" (*inukuyō*) was performed on a regular basis by women's prayer associations for safe childbirth (*koyasukō*), for instance, when a dog was found to have died in labor or when a woman died while giving birth (Kikuchi 1980). In the latter case, a memorial for the untimely dead (*nagarekanjō*) was held simultaneously by carrying a wooden spirit vessel (*futamata tōba*) to the village boundaries, to oust potential spiritual danger from the living environment.

This spirit vessel took the form of a rice ladle (*shamoji*) when placed on the grave of cats, to prevent the dead cat from becoming a haunting ghost. On the main island of Honshū of the Edo period (1604–1868), it was customary to bury a dead cat under a citrus fruit tree. On Okinawa, the preferred way of disposal was to hang the cat by its neck from a tree and let the body dry out in the wind. Every region had their way of disposing of cat remains, but there was a common need for special posthumous care for these animals (Ooki 1975).

Similarly for cats, the "Dictionary of Folkloristics" mentions that "people working in the mountains detest cats, and when they see a cat before going out to work, they will rest that day … It is unfavorable for a cat to enter the room of a person who recently passed away, lest the dead will rise… they steal the corpse from its casket, and possess people." On the other hand, female tortoiseshell cats (*mikeneko*) were believed to "prevent a ship from capsizing when taken aboard, and are highly valued by boatmen," indicating that the spiritual power of cats could work either way.

Negative connotations regarding animal death sometimes parallel the disposal of human remains. Yanagita Kunio's "Classified Folk Vocabulary of Funeral Ceremonies and Customs" (*Sōsō shūzoku goi*, Minkan Denshō no Kai 1937) explains the term "cat cemetery" (*neko zanmai*) of West Japan as "a special communal cemetery … used exclusively to bury afterbirth and dead babies." Confirming this definition, folklorist Matsuzaki Kenzō reports examples where the remains of stillborn babies and young children are buried separately from the adult graves, but together with the afterbirth and the remains of dogs and cats. One location in Tochigi prefecture had separate burial grounds behind the actual tombstones for the regular dead, children up to seven years of age, afterbirth, and dogs and cats, respectively, distinguished by the distance from the tombstone (Matsuzaki 2004).

As an urban example, the Ekō-in temple in East Tokyo possesses a death record from the 1830s to the 1850s, with posthumous names for cats and dogs. The eleven entries reported indicate that these were initiated by samurai and high-class urban citizens, and it does not appear that such posthumous care for pets had become a folk custom by then. The Ekō-in was established in 1660 for the casualties of the Meireki fires of 1657 and became known as the "temple of lost souls" (*muendera*) for taking in souls of victims of drowning, execution, and premature death. The similarity in treatment of the untimely dead and of small animals indicates that both were part of a greater category of uncanny death and spiritually dangerous entities, which explains why animal (pet) death records are found in this particular temple (Ooki 1975).

Yanagita Kunio writes: "In the village where I was born, there were about four to five 'village dogs,' but not one household actually kept a dog themselves. Their menu was unstable and their shelter was wherever they lay down" (Mame no ha to taiyō (Bean leaves and the sun), *Teihon Yanagita Kunio shū* (Standard collection of the works of Yanagita Kunio) 2, Chikuma shobō 1962). Furthermore, "before dog tax was introduced [around the 1890s, to curb rabies; note by author], most of the dogs in the countryside were village dogs without a particular owner, and because people did not interfere with their private life, they would go off to quite remote village to sow their seeds where no one knew them" (Koen zuihitsu (Essays on foxes and monkeys), *Teihon Yanagita Kunio shū* (Standard collection of the works of Yanagita Kunio) 22, Chikuma shobō 1962). It is clear that these village or community animals differ from later pet dogs and cats that "first receive their license to live when they become human property" (Totsuka 1991).

This form of ownership is shown in the way their death was handled, as well. Village women gathered to conduct *inukuyō* prayer for dead dogs and to ensure safe childbirth through prayer associations, illustrating the public character of dogs in the village. Corpse disposal sites such as *nekozanmai* and *suteba* were also community-managed, since the spiritual risk involved was a matter of concern for the entire locality. In contrast, warriors and wealthy merchants living in the urban environment were already having posthumous Buddhist names made for their individually owned pets and enshrined them in temples.

Modern animal (pet) cemeteries started to appear in Europe and the United States from the end of the nineteenth century, and Western influence has been a factor in the development of pet culture in Japan. The spread of pet cemeteries among the common people in urban environments during the 1910s and 1920s followed an increase in the number of people who started to keep animals individually just for the purpose of companionship or fun. As will become clear from the tombstones at Ooizumi Cemetery (below), dogs and cats came to be seen as personal beings worthy of individual graves. However, the influence of folk and Buddhist beliefs concerning posthumous ritual for the dead in general and animals in particular shaped modern pet rituals, as well, and signs of it can still be observed today. I look at the relationship between pre-modern and modern "pet memorial services" in two early pet cemeteries in Tokyo.

Early Pet Cemeteries and Pet Graves

Although the majority of Japanese pet cemeteries are private enterprises, some Buddhist temples manage pet cemeteries as their central activity or as a side business. Because of the strong connection in Japan between Buddhism and funerals in general, however, most private pet cemeteries are filled with Buddhist symbolism, as well.

Ooizumi Cemetery (Ooizumi reien; Nerima ward, Tokyo)

The Ooizumi Cemetery, perhaps the oldest specialized pet cemetery in Tokyo, is a branch of the Buddhist Saishinji temple in Ootsuka, Tokyo, and is located in the quiet, residential ward of Nerima. The cemetery consists of two areas where pet graves are lined up in rows, with a limited number of "special individual graves" (tokubetsu senyō bochi) near the entrance, and regular grave plots of about 60 centimeters wide (half the width of the special graves) further into the cemetery (Figure 1).

Figure 1. Special, individual pet grave (tokubetsu senyō bochi) (Ooizumi Cemetery).

This cemetery was founded in the early twentieth century when the head priest started to conduct prayer services for his beloved pet dog. In 1908, the graves were moved from the temple grounds to a separate cemetery, and in 1931 were moved to their current location. Initially pet remains were buried, but this soon changed to cremation, which is now the only option.

A "collective grave" (*kyōdō bochi*) and altar dating from 1979 is located in the middle of the cemetery. When a customer opts for this collective grave, the pet's ashes are stored together with the ashes of other pets in the collective ossuary. The collective altar for these graves is filled with cans of pet food, toys and leashes, and offerings of incense and flowers, resembling altars for the human dead with the exception of the animal-specific items. Engraved in the altar monument is the Buddhist message of this cemetery: that all living beings—humans, animals, and even plants or trees—possess innate Buddhahood (*busshō*) and should receive proper posthumous care.

At the spring and autumn equinoxes (*O-higan*), a large public memorial service is held for the pets that rest at the cemetery. The head priest recites sutras and calls out the names of customers and their deceased pets continuously for nearly thirty minutes: "Mr. A, for his beloved cat B, Mrs. C, for her beloved dog D …" This service is held in front of the "stupa for the spiritual comfort of animals" (*dōbutsu kuyōtō*), a monument under which no ashes are stored (as opposed to the individual graves), but which merely functions as a vessel for the pet souls to descend into during memorial services.

Abandoned graves are cleared periodically, but during my research in the summer of 2005 a number of tombstones from the 1910s and 1920s were still present (though not in active use any more; see Table 1). These early tombstones are shaped out of natural rocks, with the front shaved flat to accommodate an inscription. From the 1930s, the standard form becomes a rectangular pillar, following the form for human tombstones at that time. Even in this early period, the phrase *"beloved dog/cat of the house of …"* (*X-ke aibyō/aiken no haka*) is often used, and this still is a standard inscription for pet graves now. At least from the early twentieth century a custom developed in urban areas to establish tombstones for individual pets as *household members*. During this period, animals entered the spotlight with the introduction of animal protectionism from the US and Europe, and interest in war dogs grew during the 1920s and 1930s, when "dog fever" was mobilized for the national cause. Stories of loyal dogs (among them "loyal dog Hachikō," whose statue stands in front of Shibuya station) appeared in elementary school books to educate future imperial subjects on loyalty to one's master (a metaphor for the Japanese emperor; Imagawa 1996; Veldkamp 2009).

Table 1. Tombstone inscriptions—Taishō era (1912–1926).

Text	Date*	Other
Usui Hiratane aiken Bear-gō (Bear, beloved dog of Usui Hiratane)	1915	—
Kurayama-ke aichin Tama no haka (Grave of Tama, beloved Chin dog of the Kurayama family)	August 30, 1915	—
Hakushaku Sakai-ke aibyō no haka (Grave of the beloved cat of count Sakai's family)	May XX**, 1917	*Koishigawa X-harachō* (address)
Shōzen chikubyōrei (Good luck in the afterlife for the souls of my cat)	Died on January 16, 1919	
Aibyō Miiko no haka (Grave of beloved cat Miiko)	September 28, 1918	*Fukuzawa Naga-X* (name owner)
Aiken Pochi no haka (Grave of beloved dog Pochi)	1922	*Suzuki Heishirō* (name owner)

* Dates are usually written with nengō (the imperial era + year), but these have been left out of the tables in this paper for clarity.

** Xs refer to parts that have become unintelligible due to corrosion or other factors.

Reflecting this rush in popularity, from the early 1920s into the 1930s, tombstones begin to carry multiple names, such as the following example from November 1939 inscribed with "*Takano-ke aichin no haka*" (grave of the beloved Chin of the Takano family), and the names and age of death for seven dogs over a period of almost a year (see Table 2). Unfortunately, there is no way to verify the circumstances under which this tombstone was erected. Apparently, no grave was established at Maru's death in 1938, and five more Chin dogs died before the owner took that step. The worsening of living conditions following the outbreak of the Sino-Japanese war in 1937 may have been of influence, but still each of these Chin dogs was considered worthy of individual mention on the tombstone.

Table 2. Multiple names of pets on one tombstone at Ooizumi Cemetery, Tokyo (dated November 1939).

Name of Pet	Date	Age
Maru	November 7, 1938	12 years
Ekō	January 5, 1939	7 years
Kao	September 29, 1939	7 years
Chīko	October 6, 1939	11 years
Fumiko	October 11, 1939	10 years
Anpīko	October 13, 1939	8 years
Kintarō	September 29, 1939	3 years

Most individual graves at this cemetery today are dated post-1960s, and increasingly more customers have opted for the cheaper collective grave since it was constructed in 1979. Some of the grave plots have elaborate granite tombstones, while others consist of nothing more than a makeshift brick altar. Regardless of the external presence, large numbers of Buddhist grave markers (*sotoba*) decorate most graves (Figure 2). These markers are obtained through the years at periodical spirit festivals such as *O-higan* (the spring and autumn equinox) and *O-bon* (the All Souls Festival). Even though more and more people choose collective cremation instead of individual graves, many still feel the need to have memorial services performed for their pet at these occasions.

The way customers interpret the "correct" way to deal with pet remains does not necessarily follow the cemetery's Buddhist doctrine precisely. One family performed memorial services for their rabbit up to the fiftieth anniversary of its death (traditionally the concluding memorial service for a dead soul), but on the other hand, occasionally people express dislike in having their dog or cat buried out of fear for spiritual vengeance (*tatari*), reminding us of the pre-modern notion of the spirituality of cats and dogs. The notion of "pets as family" is quite common nowadays, but reality is that the dead pet may take on a special character, which will be further explored in the next example.

Jikei-in Tama Dog and Cat Cemetery
(Jikeiin-fuzoku Tama Inuneko Reien; Fuchū, Tokyo)

The Jikei-in Tama Dog and Cat Cemetery was established in 1921 and is located close to the well-known Tama Cemetery in West Tokyo. Here, individual graves are positioned around a large crematory and a three-story ossuary with a capacity of up to 100,000 urns. Most of the individual pet tombstones are from the postwar period and show a variety of sizes and forms, some mimicking human tombstones, others displaying elaborate decoration such as engraved

Figure 2. Grave markers or *sotoba* with the names of pets (Ooizumi Cemetery).

photographic images. Customers who opt for individual cremation generally rent a spot in the locker-style ossuary. There, small spaces measuring 30 centimeters on all sides contain the urn and decorations such as photographs, votive tablets (*ihai*), mini-bouquets, pet food, and toys. According to temple staff, the percentage of collective and individual cremations has evened out over the last five years, indicating a rise in individual cremations as compared with the earlier undisputed popularity of collective cremation.

The Jikei-in was established as a temple specialized in pet funeral and memorial services, but other categories of souls were accepted over the years. Prayer services for stillborn babies and aborted fetuses (*mizuko*) are performed regularly in a dedicated prayer hall, and memorial ceremonies for dolls (*ningyō kuyō*) are held annually. A general monument for animal souls was added in 1989, one for laboratory animals was added in 1992, and in 2000, a monument for animals kept in schools was created. The remains from collective cremations are scattered in a collective grave, which was established in the early 2000s in the form of a small bamboo forest.

Particular to the Jikei-in cemetery is the Hall of Great Compassion (*Daihiden*; est. 1972), where the remains of pet owners are stored together with the remains of their pet(s). A cemetery pamphlet describes this hall as an "ossuary for people with no one to pray for their afterlife and those afraid of becoming wandering spirits (*muen*), to share their journey to the hereafter with the animals that accompanied them in their loneliness." On each floor, a Buddhist altar (*butsudan*) and individual locker-style ossuaries holding votive tablets for the souls of pets and their masters can be found (see Table 3).

In examples 1 and 2 (Table 3), we see that the pet, after having "accompanied … [their owners; note by author] in the loneliness" of their elderly years, is either enshrined before the owner's death, or added after the owner has passed away, perhaps by a son or daughter. Some of the tablets are not dated, such as in example 3, in which case the owner added the names of pets from a distant past when the altar was first taken into use (Figure 3). In examples

Table 3. Joint enshrinement of votive tablets for owner and pet remains (Daihiden, Jikei-in Tama Dog and Cat Cemetery, Tokyo).

Text on Votive Tablet (*ihai*)	Date
1) *Seiun Masasue Zenjōmon'i* (posthumous Buddhist name, male)	February 23, 1990
a) *Aiken Pepe no reii* (soul of beloved dog Pepe)	May 21, 1981
b) *Aiken Mīmi no reii* (soul of beloved dog Mīmi)	June 8, 1991
2) *Jōkō Myōyō Taishi'i* (posthumous Buddhist name, female)	March 19, 2003
a) *Aibyō Chibi/Mami no reii* (souls of beloved cats Chibi and Mami)	October 3, 1989/March 30, 1991
b) *Aibyō Chīko/Tora no reii* (souls of beloved cats Chīko and Tora)	March 9, 2003/October 24, 2003
c) *Aibyō Rirī/Shiro no reii, aibyō Yamamoto* (souls of beloved cats Rirī, Shiro, Yamamoto Chīko, Kuro and Mimi)	—
3) *Motohashi Maria/Motohashi Jūichi/Kācha/Arekisē reii* (souls of Motohashi Maria, Jūichi, Katya, Alexei)	—
a) *Chibi/Chibi/Zerochika reii* (souls of Chibi, Chibi and Zirochka)	—

Figure 3. Votive tablets for humans and pets together (Jikei-in Tama Dog and Cat Cemetery).

1 and 2 (Table 3) there is a votive tablet for only one person, suggesting that the only "family members" in the last days of their life were pets. Such was the case with Russian-born Maria (example 3, Table 3), who immigrated to Manchuria after the Russian revolution and met and married a Japanese journalist there. After the war, her husband was repatriated to Japan, but by the time she finally arrived in Japan in 1953, her husband had already passed away. In her

final years, "communication between her and the world around her had shifted from people to cats" (Asahi Shimbun, January 3, 1989).

It is interesting to see that joint enshrinement of non-human souls as "family members" is not limited to pets: on some of the altars, votive tablets for deceased babies, miscarriages, and aborted fetuses (*mizuko*) are encountered, as well (see Table 4). In example 1 (Table 4), the souls of a woman and her baby are enshrined together, and in example 2 there is no tablet for a deceased adult yet, but the baby and the "beloved dogs and cats" have already been put to rest. To find the souls of pet animals and dead babies resting in the same ritual space bears resemblance to the pre-modern "disposal grounds" (*suteba*) or the "cat cemetery" (*nekozanmai*) mentioned above. Animal and baby deaths were both considered untimely or abnormal deaths (*ijōshi*), and in the last thirty years both have reappeared as objects for posthumous worship. We may say that pets and *mizuko* bear similarity in that they have come to be included in the modern version of the Japanese "extended family." Furthermore, neither of the pet cemeteries openly mentions the importance of memorial services to appease the souls of the dead in relation to spiritual vengeance (*tatari*). Except for a small percentage of customers who are skeptical about burying animal remains, the majority of pet owners nowadays consider their dogs and cats to be intimate creatures instead of spiritual beings, and often insist on taking the remains home.

Table 4. Joint enshrinement of votive tablets for persons and mizuko (Daihiden, Jikei-in Tama Dog and Cat Cemetery, Tokyo).

Text on Votive Tablet (ihai)	Date
1) *XX-Myōtoku Seitaishi* (posthumous Buddhist name)	2004
a) *Sōsei Kōro Eiji no reii* (soul of a prematurely deceased baby)	April 28, 1955
2) *Egyokuzen Eiji* (posthumous Buddhist name, baby)	November 23, 1962
a) *Aikenbyō no shoreii* (souls of beloved dogs and cats)	October 26, 1973

This change in pet owners' consciousness (especially in urban areas) received further impetus during the rapid economic growth from the 1960s, bringing forth an increase in the number of temples and private companies that provided pet funeral services. The limited spatial conditions of the urban environment caused many cemeteries to build compact, high-volume, locker-style ossuaries during this time. The Jikei-in apparently already had a small ossuary during the 1940s, but the current three-story ossuary dates from 1963. The Ekō-in in Ryōgoku, known for its animal memorial services from the mid-1800s, built a Merit Hall for One Million Animals (*Hyakumantō ekōdō*) in 1962 and expanded this in 1972. There is no locker-style ossuary at the Ooizumi Cemetery, but a collective grave was established to deal with increasing demand in 1979.

As we saw earlier, during the early twentieth century two patterns of tombstone inscriptions emerged: the generic "*X-ke aiken/aibyō no haka*" ("Grave for the beloved dog(s) and cat(s) of family X"), which may signify any number of "family animals" buried, and the more specific "*X-ke aiken/aibyō* [name] *no haka*" type for individual animals. The custom to inscribe names of successive pets into one tombstone eventually led to the installation of a separate plaque (*boshi*) next to the tombstone, just for this purpose in the postwar period.

The option of a joint final resting place for owner and pet is still not common among pet cemeteries, but recently, regular cemeteries have been taking steps to bridge the divide. Several cemeteries in the Tokyo area have a *wizupetto* section (i.e., "with pet," an adjective that points at various products that accommodate a joint life experience for owner and pet, such as apartments). Besides regular plots for the human dead, these cemeteries provide grave plots that allow the storage of human and animal ashes together, and accommodate "a new style of memorial services [*kuyō*; note by author] in which your pet is cared for in eternity together with the family" (<http://www.ohnoya.co.jp/cemetery/with.shtml> Accessed August 7, 2008).

Japanese Folklore and the Politics of Memorial Services (*kuyō*)

In Japan, the purpose of memorial services for the human dead is to guide the newly dead soul towards ancestorhood, by performing periodic anniversary rituals (*kaiki*). Upon conclusion of these rituals in the thirty-third year or fiftieth year after death, the soul sheds its individual character and becomes an anonymous "ancestor spirit" (*tomurai-age*; Iwamoto 1999). These rituals also provide the bereaved family a way to digest grief and loneliness, and we may say that they are aimed at the living just as much as they are at the dead.

How does this work in the case of pets? When dogs and cats were considered spiritual beings within the village, they received treatment similar to the untimely human dead. In contrast, modern temples and private pet cemeteries decide upon the form of pet memorial services one-sidedly, and customers consume these forms as a commercial service. Still, varying interpretations by these customers suggest that the system of posthumous ritual for pets is both flexible and adaptive in its nature.

The commercialization of rituals for the non-human dead has been studied in the case of "*mizuko kuyō*," memorial services for aborted fetuses. During the period that pet funerals started to increase (the 1970s), "ritual manuals" concerning the proper way to worship ancestors and to deal with spiritual trouble and vengeance appeared as well, and *mizuko kuyō* played a central role. According to one researcher, the formation process for *mizuko* first grabbed "the hearts of housewives, who had abortions and miscarriages in the turbulent postwar years" during the early 1970s, and was revived later when "*mizuko* songs and *mizuko* characters were created and forced themselves upon young women," during the late 1970s to the mid-1980s. This movement eventually settled in as a "custom of the Shōwa era,"[2] covering up the fact that support for *mizuko* veneration had come from mental strategies and pressure by editorial staff at women's magazines (who ran frightening stories about baby spirits) and by religious entrepreneurs (mostly men), indicating that power relations and gender were significant factors in this commercial spiritualism (Morikuri 1995).

Weekly magazines and other publications do not seem to have had the same significance in the case of pet death. The advice in the few existing ritual manuals is so scattered that it would be hard to ascribe any influence to them, and the temples in this article do not show signs of pressure or mind play. Rather, pet cemeteries revolve around an owner-centered narrative which primarily stresses the "pitifulness" (*kawaisō*) of dead pets. For example, a middle-aged lady at one of the cemeteries confessed that she had bought a grave for two cats that had lived with her for more than a decade, but that she had the name of her adult daughter and son engraved in the tombstone out of concern that the grave would be abandoned after her death. There is no "required course" for memorial services set by the temple (only suggested funeral packages), and most owners have the services performed in a way they themselves see fit. This strengthens the impression that these pet funerals and memorial services are now primarily about soothing the bereaved, and less about belief in animal souls.

It is the owner's choice to have their pet cremated or to perform posthumous ritual, but the circumstances under which this choice is made differ greatly between rural and urban environments. In recent years, animal hospitals and local authorities have been known to recommend pet cemeteries, and there are few other options in urban areas. In rural areas, this problem is often still solved within the locality. The head priest of a temple on the island of Sado in Niigata prefecture explained that people occasionally bring their dead pet and ask him to write a sutra on a piece of paper, roll it into the dog's ear, and conduct a Buddhist prayer for safe afterlife. Most of these pets are then buried in the hills outside the village, creating a harmony between the decay of the pet's body and the fading away of memories and grief. In an urban environment, such a way of disposal is limited by spatial conditions and hygienic regulations, narrowing the choice down to disposing the remains as trash, or going to a commercial cemetery.

It has been pointed out that posthumous care for pets tends to be relatively short, despite that a central role as "child" in the family is often attributed to it during its lifetime. In 1988, at the Dog and Cat Graveyard of Takao Cemetery (est. 1975) in Western Tokyo, 1230 of the 1343 cremations were collective cremations. Out of 59 individual graves created that year, 39 were abandoned within four years, illustrating a short span of commitment between the owner and the dead pet (Takeda 1994).

Nevertheless, attention to the grave varies from person to person; from owners who attend the cremation but never visit again, to owners who take care of the pet grave for decades after. Pet cemeteries do not force the customer in any direction, and posthumous care for pets relies to a large extent on the owner's individual effort. Although belief in spiritual vengeance by animal spirits is becoming less urgent for many people, its traces can still be noted. Even so, the significance of memorial services hinges on the circumstances of the bereaved more than ever.

Concluding Remarks: Animals Becoming Pets, and Human Convenience

As mentioned above, from the early twentieth century onward, significant change has occurred in the way Japanese people live together with animals. An important factor has been the emergence of private pet ownership among urban citizens, creating an environment for the modern owner–pet relationship of intimacy. Because of this, people's traditional views of animals have slowly been pushed to the background; now the owner–pet relationship is more related to owner convenience. Additionally, memorial services for small animals as spiritual beings have lost their local significance and are now a matter of supply (temple, cemetery) and demand (pet owner), thereby also losing their consistency with the ritual system for untimely dead.

Pets are entirely dependent on their human owners, and their position as "family" is supposedly reinforced by, and illustrates changes in, the postwar Japanese family. There are plenty of stories about elderly couples who purchase a pet to fill the emptiness left after their children leave the home, or single households where the cat becomes the owner's best friend. On a critical note, Totsuka Hiromi points out that the image of the "pet as family" is a narrative in which pets function as a "device to reinforce the narrative of a diluted 'family,' " and that pet-related goods and services (including pet cemeteries) function as the "hardware needed to guarantee the narrative of the 'pet as family'" (Totsuka 1991).

Changes in the Japanese household have included the emergence of the nuclear family that replaced the extended family household of the *ie*[3], and subsequent development into post-nuclear forms such as single households and husband–wife households. Hereby, a favorable environment is created for the pet animal to thrive as a remedy for loneliness and a stimulant for household life. The number of human members in the Japanese household may have decreased over the years, but a re-expansion into non-human entities such as pets and perhaps even *mizuko* shows us a new household structure that is based on the relationships of each of its individual members, rather than those of the whole group.

Seen from the perspective of pet funerals and memorial services, the "pets as family" expression appears to be valid, in particular when the pet is alive and visible. Individual graves in the above-mentioned pet cemeteries resemble the human grave to a large extent, but a distinct contrast remains between the pet that lived together as a "family member" and the reality of separate resting places for owner and pet after they have passed away. This illustrates that there are limits to the "familiar" aspect of pets: it may only last until they die.

The intimacy a pet owner had with his/her pet during a lifetime tends to turn into a matter of convenience after the initial shock of loss, when long-term posthumous care becomes involved. Pet cemeteries in Japan are one example of a pet industry that is highly based on the market principle, but comprises deeper (traditional) meanings that cannot be rationalized. These meanings also influence the consciousness of pet owners when they perform memorial services. A small number of people choose not to follow the package of prayer services most temples and cemeteries offer and are aware of the traditional meaning of animal death, regardless of whether it was their lifetime companion or not.

Communal animal memorial ceremonies of yore have now been replaced by individual rituals for privately owned pets, and the meaning of animal burial has shifted from solace for the animal spirit to comfort for the bereaved human being. In spite of looking very similar on the outside, the spiritual care processes of human death and pet death are quite different in reality. Therefore, it remains to be seen if pets ever come to receive truly "humane" (*ichininmae*) posthumous treatment.

Notes

1. For an overview of different varieties of animal memorial services in modern Japan, see Veldkamp (2005).
2. The reign of emperor Shōwa (1926–1989), known elsewhere in the world by his first name Hirohito, was a turbulent period for modern Japan because it ranges from the ultranationalistic period up to 1945 to the postwar revival of economic growth in Japan during the 1960s and 1970s, to the economic bubble of the late 1980s. Many aspects of "Japaneseness" in modern Japanese identity were settled in developments during this period.
3. The term *ie* refers to the traditional Japanese household made up of an extended family of blood relatives (grandparents, their son and his wife and their children), but also maids, farmhands, and others not related by blood, all living on the same estate or in the same house.

References

Imagawa, I. 1996. *Inu no Gendaishi* (Modern History of Dogs). Tokyo: Gendai shokan.
Iwamoto, M. 1999. Shinibasho to Kakugo (Places to die and preparedness). In *Minzokugaku no Bōken 4— Kakugo to Ikikata* (Adventures in Folkloristics 4—Preparedness and Ways of Life), 172–214, ed. M. Iwamoto. Tokyo: Chikuma shobō.
Kikuchi, K. 1980. Tonegawa karyūiki no ikunuyō (Dog memorial services in the downstream area of Tone river). *Nihon Bukkyō* (Japanese Buddhism). Vol. 50–51. Nihon bukkyōshi kenkyūkai (Japanese association for research on Buddhist history).

Matsuzaki, K. 2004. *Gendai Kuyō Ronkō—Hito / Mono / Dōshokubutsu no Irei* (Treatise on Modern Memorial Services—Comforting the Souls of People, Objects, Animals and Plants). Tokyo: Keiyūsha.

Minzokugaku, K. ed. 1952. *Minzokugaku Jiten* (Dictionary of Folkloristics). Tokyo: Tokyodō.

Morikuri, S. 1995. *Fushigitani no Kodomotachi* (Children from the Valley of Wonder). Tokyo: Shinjin ōraisha.

Ooki, T. 1975. *Neko no Minzoku* (Cat Folklore). Tokyo: Tabata shobō.

Takeda, M. 1994. Gendai Nihon no kazoku to pettokuyō—shiekidōbutsukan to dōbutsukuyō no hensen (Pet memorials and the modern Japanese family—changes in animal memorials and views on work animals). *Shūkyōgaku Nenpō* (Annual report on religious studies) 24.

Totsuka, H. Petto to kazoku no monogatari (The narrative of pets and family). In *Shirī zu Henbō suru Kazoku 4—Kazoku no Fōkuroa* (The Changing Family (series) 4—Folklore of the Family), 251–270, ed. C. Ueno et al. Tokyo: Iwanami shoten.

Veldkamp, E. 2005. Memorializing animals in Japan. In *Social Memory in Japan*, 58–74, ed. Tsu Yun Hui. Kent, UK: Global Oriental Press.

Veldkamp, E. 2009. Eiyū ni natta inutachi—gunyōkenirei to dōbutsukuyō no henyō (Dogs that became heroes—war dog commemoration and change in animal memorial services). In *Hito to Dōbutsu no Nihonshi 3—Dōbutsu to Gendaishakai* (Japanese History of Man and Animal 3—Animals and Contemporary Society), 44–68, ed. Y. Suga. Tokyo: Yoshikawa Kōbunkan.

Yamada, M. 2004. *Kazoku Petto—Yasuragu Aite wa, Anata Dake* (The Family Pet—Only You Make me Feel at Ease). Tokyo: Sanmāku shuppan.

Reasons for Relinquishment and Return of Domestic Cats (*Felis Silvestris Catus*) to Rescue Shelters in the UK

Rachel A. Casey*, Sylvia Vandenbussche[†],
John W.S. Bradshaw* and Margaret A. Roberts[‡]

*Department of Clinical Veterinary Science, University of Bristol, UK
[†] Beeldhouwersstraat 46, Antwerp, Belgium
[‡] Cats Protection, National Cat Centre, Sussex, UK

Address for correspondence:
Rachel Casey, Department of
Clinical Veterinary Science,
University of Bristol,
Langford, Bristol
BS40 5DU, UK.
E-mail:
Rachel.Casey@bristol.ac.uk

ABSTRACT Significant numbers of cats enter rescue and re-homing facilities each year, over half of which are relinquished directly by owners. Identifying the reasons why owners decide to give up their pet is an important step in the development of education strategies to encourage retention of cats by their owners. In addition, identifying why adopting owners fail to retain their new cats is important in the refinement of homing policies. Characteristics of 6,089 cats relinquished and returned to 11 rescue facilities in the UK were recorded over a year. In addition, information was collected on the reason why owners gave up, or brought back, their pet. Sixty percent of cats and kittens entering shelters were relinquished by owners, with 19% being due to owner circumstances, such as moving to rented accommodation or changes in family circumstances. Seven percent were for behavioral reasons, and 5% because of the occurrence of allergy or asthma in owners. Returned cats were significantly more likely to be older (Mann Whitney U, $Z = -9.167$, $p < 0.001$) and neutered (Pearson $\chi^2 = 110.0$, $df = 2$, $p < 0.001$) than the general relinquished population. The reasons for original relinquishment and return of owned cats were also significantly different (Pearson $\chi^2 = 84.4$, $df = 6$, $p < 0.001$), with 38% of cats being returned for behavioral reasons, and 18% because of allergy or asthma. The commonest behavioral reason for both relinquishment and return was aggression between cats in the household.

Keywords: domestic cat, relinquishment, rescue, welfare

Relinquishment of Cats to Rescue Centers

 It is a significant concern for those interested in feline welfare that large numbers of domestic cats are relinquished into rescue shelters each year. In addition, a proportion of cats that are homed from shelters do not remain in their new homes. Identifying reasons why owners give up their cats, or fail to retain them after adoption, is an important first step in directing further research and educative programs aimed at reducing the numbers of cats needing to enter the shelter system. In the UK, significant numbers of cats are relinquished into rescue organizations each year: for example, in 2001, 6,089 were relinquished to the adoption centers of the UK cat charity Cats Protection. Although Cats Protection has a non-euthanasia policy, overall many cats are euthanized in rescue organizations because of the huge numbers of unwanted pets in the population. In fact, euthanasia of unwanted animals is thought to be the leading cause of death for cats in the USA (Olson et al. 1991), with approximately 4 million being euthanized annually (Clifton 1993, cited in Patronek et al. 1996). The 16 "open admission" shelters in Massachusetts, USA, admitted nearly 56,000 cats in 1993, and almost 41,000 (73%) of these were euthanized (Luke 1996).

Patronek and colleagues (1996) investigated risk factors for relinquishment of cats to a rescue shelter by comparing 218 owners who relinquished cats to shelters with a control group of 459 cat owners who had not relinquished their pets, recruited through random digit dialing. A number of factors were found to be associated with increased risk of relinquishment in this study, based on univariate analysis. However, in the final multivariate model, the factors remaining (in order of strength of association with relinquishment) were: inappropriate elimination; specific expectations about a cat's role in the household; inappropriate care expectations (strongly associated with occurrence of problem behaviors); allowing the cat outside; the cat being sexually intact, and owners never having read a book on cat behavior.

Luke (1996) reported on the reasons for cats being relinquished into shelters in Massachusetts. Of these, 24% were owned kittens and 18% were unowned kittens; 17% were stray adult cats; 10% due to owners moving/landlord problems; 8% were due to behavioral problems; 7% were because the owner requested euthanasia of the cat; 7% because the owner just no longer wanted the cat; 6% were for financial reasons, and 4% because of owner allergies (Luke 1996).

In order to acquire a more detailed picture of why animals are relinquished, DiGiacomo, Arluke and Patronek (1998) interviewed people relinquishing a pet to a shelter in Boston, Massachusetts in 1997, using a semi-structured, open-ended format. The behavior of the pet was stated to be a cause of relinquishment in 32% of respondents, followed by allergies (18%), moving house (18%), and renting/landlord problems (8%). The remaining seven people interviewed expressed some variation on not being able to keep their pet any longer, for example, because of overcrowding. Although the species of pet relinquished was not identified, and the sample size was small (38 respondents), this study did highlight that, over the course of many interviews, it was rarely a single problem that led to relinquishment; rather, it was a combination of factors. For example, one owner who first stated "allergies" as a reason for relinquishing her cats had actually put up with them causing asthma to her grandchild for five years, and the factor that had precipitated relinquishment was the development of inappropriate toileting in one of the cats (DiGiacomo, Arluke and Patronek 1998).

Salman and colleagues (1998) interviewed 3,772 owners relinquishing pets to 12 shelters distributed across the USA; 1,409 of relinquishing owners brought cats or kittens. Only owned animals whose owners were willing/able to be interviewed were included in this study (13% of the intake to the shelters was from strays). The many individual reasons for relinquishment were condensed into 12 broad categories. Overall, the categories involving behavioral reasons (aggression towards people, aggression towards other animals, and other behaviors) made up

33% of relinquishments, which was the second most common reason after "human lifestyle" (35%). Of the behavioral reasons, the most common were house-soiling (5%); incompatibility with other pets (2%); destruction inside the house (2%); aggression towards people (2%); aggression towards animals (1%); biting (1%); not friendly (1%), and "too active" (1%).

In addition to asking about the reason for relinquishment, Salman et al. (1998) also asked owners relinquishing pets the relative frequency (on a four-point scale) at which common behavioral problems (on a list provided) occurred in their pet in the month prior to relinquishment. When the responses "sometimes," "mostly" and "always" were combined, 23.6% of cats soiled in the house at some point; 24% did some damage to the house; 33.1% were hyperactive; 42.6% were too noisy; 43.5% acted fearful; 13.3% had growled at people; 12.5% had attacked an animal; 10.8% had escaped; 8.9% had bitten someone, and 16% had scratched someone. Owners were also asked a series of questions to investigate their general understanding of animal behavior, and some interesting misconceptions were revealed. For example, 58.3% of cat owners thought that "animals misbehave out of spite," and 29% thought that rubbing their pet's nose in its mess was an effective discipline (Salman et al. 1998).

In a smaller but similar study conducted in a single shelter in Ohio, 47% of cats relinquished had come from private owners, and more than two thirds of these were less than two years of age (Miller et al. 1996). The most common problem behaviors in cats in this study were scratching furniture and aggressive play. Miller and colleagues (1996) also found that relinquishing owners commonly had unrealistic expectations of their pet and a poor understanding of feline behavior. As in Salman et al. (1998) and Patronek et al. (1996), many owners reported that moving house, or moving into rented accommodation where pets were not allowed, was a reason for relinquishment.

In summary, the reasons for relinquishment of cats to shelters appear to be complex. There is obviously a large problem of "un-owned" cats and unwanted kittens entering shelters (Luke 1996). Previous studies suggest that the owner's lifestyle or circumstances appear to be an important cause of relinquishment of cats; for example, when they move to accommodation where pets are not welcome (Bailey 1992; Patronek et al. 1996; Salman et al. 1998). Owners having a misplaced expectation about their pet's role in the family can also be a factor in the decision to relinquish (Salman et al. 1998; New et al. 2000). Many owners have a poor understanding of cat behavior (Miller et al. 1996; Salman et al. 1998; New et al. 2000), and even having read a book on cats appears to have a protective effect against relinquishment (Patronek et al 1996). In addition, factors associated with the animals themselves are important: a younger age, being sexually intact, and being mixed breed appeared to be risk factors for relinquishment (Miller et al. 1996; Patronek et al. 1996; Salman et al. 1998). The occurrence of certain behavioral problems in cats, such as house-soiling, has also been found to be an important factor in owners' decisions to relinquish them to shelters (Miller et al. 1996; Patronek et al. 1996; DiGiacomo, Arluke and Patronek 1998; Salman et al. 1998).

Return of Cats After Adoption

Neidhart and Boyd (2002) conducted a survey to investigate the extent to which adoption of dogs and cats from rescue organizations was successful. Most owners (94%) reported that they were satisfied with their new cat. However, 20% of all adopting owners (including those who adopted a dog) had relinquished their pet again by one year after adoption: 4% of these in the first two weeks, and another 8% by six months post adoption. Failure to retain a pet was higher where the pet was over a year old at time of adoption, where the pet had been acquired for a child, and in households where income was less than USD35,000. The most common reason for no longer having a pet was death of the animal: this was higher in cats (36%) than dogs (15%). Twenty percent of cats returned were due to the occurrence of behavioral problems, broken down into "Did not get along with other animals" (5%), "Did not get along with children" (2%), "Did not get along with adults" (2%), "Other compatibility issues" (2%), "Destructive" (2%), "House-soiling" (2%), and "Other behaviors" (5%). The other reasons cited for return were the cat running away (17%), the cat having a medical problem (7%), and the owners having an allergy to the cat (10%).

Aim of this Study

The aim of this study was to use the data routinely collected by shelter staff when cats are relinquished into rescue centers, in order to provide further insight into the reasons why owners give up their cats to rescue centers, and why they return them to centers after homing. This was done in order to inform the direction of further research and educative programs.

Methods

Data were collected from 11 Cats Protection re-homing centers in England and Wales. These centers were all cat-only facilities and were made up of five larger centers run centrally by the charity and six smaller centers managed at a local level. Two centers were based within built up areas (one suburban and the other in an industrial urban area), and the remainder were located in rural areas. Over a 12-month period (January to December 2001), owners relinquishing cats filled in a standard "relinquishment form." This short questionnaire obtained data on the age, breed, color, sex, and neuter status of any relinquished cat, and asked owners to indicate if they were returning a cat that they had obtained from the shelter. It also had a "free answer" section, where owners were asked for the reasons that they were relinquishing their cat. The individual responses given in this section were subsequently grouped into nine categories: "Owner circumstances," "Allergy/asthma," "Behavioral reason," "Unwanted kittens," "Multiple factors/Can't cope," "Owner pregnancy/young child," "Too many cats in household," "Cat pregnant or unwell," and "No reason given/other reason." The category "Owner circumstances" included the death of the owner, moving house, living in rental accommodation, financial circumstances, problems with neighbors, divorce or family breakdown, and owner illness. Un-owned cats that came into shelters were categorized as "Stray/Abandoned," or "Transferred from other facility." The latter description was applied to cats that had arrived from another shelter, volunteer carer, or charity. Most of the data analyses in the study were descriptive. Characteristics of relinquished cats that had previously been homed from the shelter were compared with those that had not, using the chi-square test (SPSS for Windows 12.0, Chicago, IL, USA).

Results

Relinquishment of Cats

In total, records were obtained for 6089 cats relinquished to the 11 centers. Their ages ranged from 0 to 20 years (Figure 1), with 2322 (38%) being kittens (less than 4 months), 3157 (52%) being adults (4 months to 7 years) and 610 (10%) being older cats (over 7 years). Forty-five percent (n = 2752) were males and 52% (n = 3150) were females, with no record of sex being reported for the remaining 3%. Forty percent (n = 2432) were neutered and 59% (n = 3608) were not, reflecting the high number of kittens in the relinquished population. When only cats over 6 months of age were included, 68% were neutered (n = 2289). Domestic Short Hairs

(DSH) made up 87% of the cats, 10% were Domestic Long Hairs (DLH), 1% were "Persian type," 0.4% were "Oriental type," and 0.9% were other breed types. Information on breed was missing in 16 (0.3%) cats.

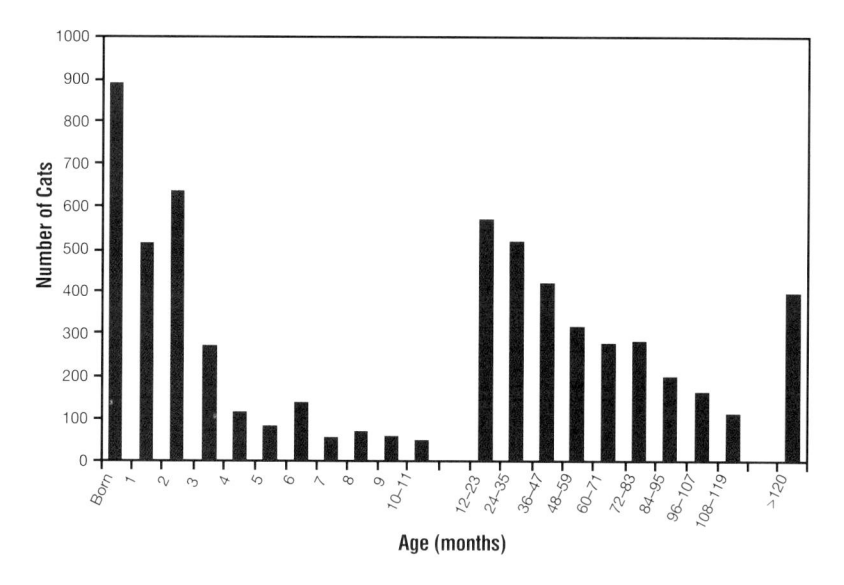

Figure 1. Distribution of ages of cats relinquished to rescue shelters.

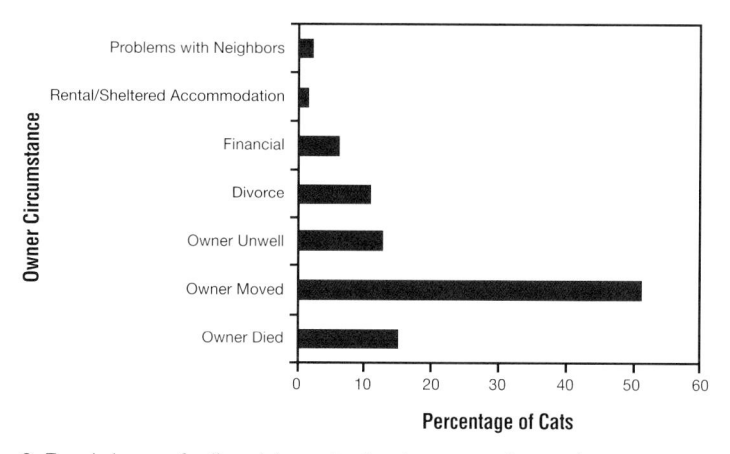

Figure 2. Breakdown of relinquishments due to owner circumstances.

The most common reasons given for the cats to be relinquished into the centers were: found abandoned or straying, owner circumstances (see Figure 2), unwanted kittens, transfer from other facilities, and allergy/asthma, followed by behavioral reasons at 7% of the total (Table 1). Out of the 449 cats relinquished for behavioral reasons, 161 (36%) owners reported aggression between cats in the household, 77 (17%) reported house-soiling, 65 (14%) reported aggression towards people, 53 (12%) reported fearful behavior in one or more contexts, 45 (10%) reported problems associated with dogs, 44 (10%) reported one or more other behaviors, and 4 (1%) reported scratching (Figure 3).

Reasons for Relinquishment and Return of Domestic Cats...

Table 1. Reasons for relinquishment of cats to shelters.

Reason	Number of Cats	Percentage of Cats
Stray/Abandoned	1920	31
Owner Circumstances	1175	19
Transfer from Other Facilities	569	9
Allergy/Asthma	280	5
Behavioral Reason	449	7
Unwanted Kittens	864	14
Multiple Factors ("Can't Cope")	295	5
Pregnancy/Young Child	130	2
Too Many Cats	133	2
Cat Pregnant or Unwell	75	1
No Reason Given	199	3

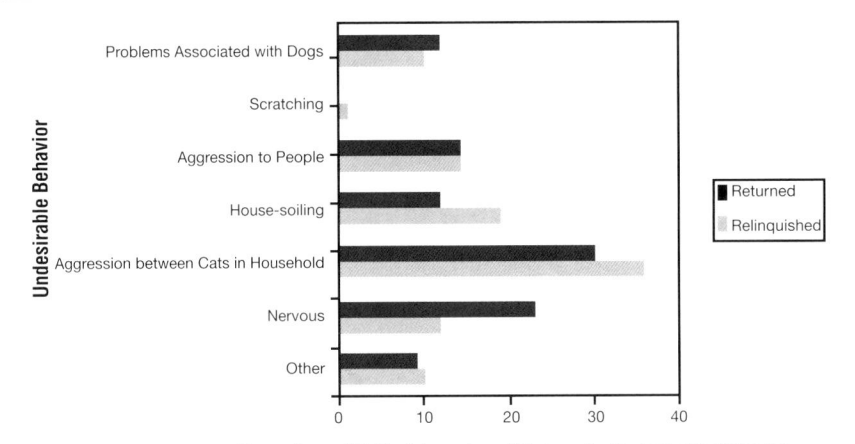

Figure 3. Behavioral reasons for the relinquishment and return of cats to shelters.

Return of Previously Homed Cats

Of the 6089 relinquished cats, 166 (3%) had been returned to the shelter from which they had been originally homed. However, this figure is unlikely to be representative of the proportion of returned cats, since this information on whether cats were "returns" or not was missing in 3162 (52%) cases. In the 166 known returns, the age range was from 0 to 19 years, with a mean of 4.4 years. Seventy (42%) were males, and 95 (58%) females. Seventy-nine percent were neutered and 20% were not, and no information on neuter status was available for 2 cats (1%). Eighty-three percent were DSH, 12% were DLH, 2% were "Persian type," and 2% were other breed types. There were no "Oriental types" in the returned population. The main reason given for the return of these 166 cats was behavior (38%: Table 2), a far greater proportion than for relinquishments in general (Figure 4). When reasons for surrendering the cat that were unlikely to apply to returns, that is, "too many cats," "transfer," "unwanted kittens," and "strays," were excluded, significant differences remained between the reasons for the surrender of relinquished and returned cats (Pearson $\chi^2 = 84.4$, $df = 6$, $p < 0.001$). Apart from behavior, the greatest difference between the two populations was a higher proportion of returns for allergy/asthma. This may partly be explained by a previously unsuspected allergic reaction being triggered in a family member by their first adoption of a cat.

Table 2. Reasons for return of cats to shelters.

Reason	Number of Cats	Percentage of Cats
Stray/Abandoned	8	5
Owner Circumstances	38	23
Transfer from Other Facilities	3	2
Allergy/Asthma	30	18
Behavioral Reason	63	38
Multiple Factors ("Can't Cope")	6	4
Pregnancy/Young Child	6	4
Cat Pregnant or Unwell	3	2
No Reason Given	3	2

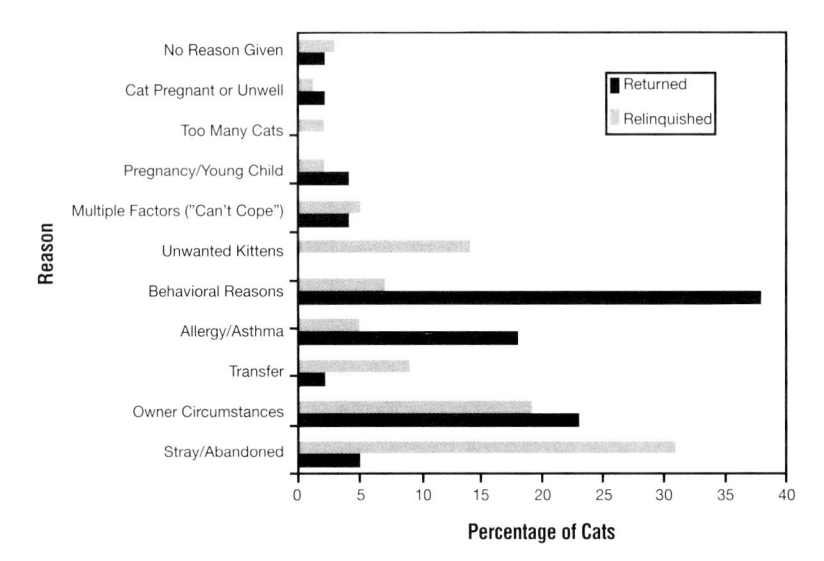

Figure 4. Reasons for owner relinquishment of cats to shelters and reasons for return to shelters after homing.

The returned cats were significantly older than the general relinquished population (Mann Whitney U, $Z = -9.167$, $p < 0.001$). There was no significant difference in the gender (Pearson $\chi^2 = 1.21$, $df = 1$, $p = 0.27$) or breed type (Pearson $\chi^2 = 5.00$, $df = 4$, $p = 0.29$) of returned cats compared with the general relinquished population. However, the returned population was significantly more likely to have been neutered (Pearson $\chi^2 = 110.0$, $df = 2$, $p < 0.001$). Of those returned for behavioral reasons, 21 (30%) had been as a result of aggression with other cats in the household, 16 (23%) because of fearful behaviors in a range of contexts, 10 (14%) because of aggression towards people, 8 (12%) because of house-soiling, 8 (12%) because of problems associated with dogs, and 6 (9%) for other reasons (Figure 3). There was little difference in the profile of behavioral reasons for returned versus relinquished cats (Pearson $\chi^2 = 8.30$, $df = 6$, $p = 0.22$).

Discussion

Comparison with Results of Previous Studies

A significant proportion of cats relinquished to shelters in this study were abandoned or stray animals and unwanted kittens, as was also reported by Luke (1996) and Miller et al. (1996). The number of stray/abandoned animals was higher, however, than in the population reported by Salman et al. (1998) (31% compared with 13%). It is difficult to know why there is this difference, although it is likely to reflect the proportion of feral/stray cats in the vicinity of shelters, socio-economic factors in the area around shelters, and the policy of individual shelters as to the procedures for dealing with feral/stray cats (e.g., attempting to re-home, or neuter and return). The results of this survey also support the findings of Patronek et al. (1996), in that cats under 6 months of age, DSHs, and entire cats have a relatively high prevalence in the relinquished population.

Of the cats relinquished by owners, the reasons for relinquishment broadly concur with the findings from previous studies, with changes in owner circumstances, the occurrence of allergies or asthma, and the occurrence of undesirable behaviors in the cat being commonly cited. In the general population, the prevalence of allergy to cats is from 8 to 10% (Foucard 1991; Shabaan et al. 2008), making the 5% of owners citing this as a reason for relinquishment in our study not unusual.

One noticeable difference in our study is the low number of relinquishing owners stating "financial reasons" to be the main factor in the relinquishment of their cat (75 owners, or 1% of the sample, gave this reason). In contrast, Luke (1996) found 6% of owners gave this reason, and Patronek et al. (1996) found family income to be a significant risk factor for relinquishment. This difference may reflect the socio-economic areas in which the shelters are located in the different studies, or cultural differences between the UK and USA in willingness to discuss issues related to finances. In addition, the "open" format of the questionnaire in this study may have led to owners choosing to report the least embarrassing or sensitive reason, out of several co-occurring factors, when completing the questionnaire, as was found by DiGiacomo, Arluke and Patronek (1998).

The 7% of cats relinquished for behavioral reasons in this study is comparable with the 8% reported by Luke (1996). However, Salman et al. (1998) reported that 33% of relinquishments were for behavioral reasons. Since Salman et al. (1998) only included cats relinquished from owners, the stray or abandoned cats should be excluded from this population, for a valid comparison. However, the proportion reported by Salman et al. (1998) is still higher than the 16% of "non stray" cats found in our sample. One possible reason for the difference in these samples may be related to the euthanasia policy of the shelters involved. In our study, all the shelters had a non-euthanasia policy. This potentially could influence owner reluctance to mention any behavioral problems in either direction. Being less concerned about potential euthanasia of their pet may make owners more likely to mention the occurrence of behavioral problems. However, non-euthanasia shelters may be perceived by owners as being more selective about which cats they take in, and hence behavioral problems are not mentioned, to reduce the risk of their pet being turned away. Although the policy of CP shelters is to take cats in from waiting lists in a chronological order, irrespective of the reason for relinquishment, owners were not necessarily aware of this, and may have been reticent in mentioning anything that might decrease the chance that the shelter would take their cat. In the study by Salman et al. (1998), relinquished animals not suitable for re-homing were euthanized (55% of the population had this fate), and this may have led to owners being less likely to conceal behavioral reasons for relinquishment, as identified by DiGiacomo, Arluke and Patronek (1998).

The individual behaviors shown by cats in this study which led to relinquishment are also broadly similar to those shown in previous studies. The behaviors which led owners to relinquish their cat appear to at least partially depend on the type of behavior shown. Those which impinge on the home environment or lifestyle of owners, such as house-soiling and aggression, are often perceived to be more of a problem and also lead to owners seeking professional help with the behavior of their cat (Casey and Bradshaw 2001), as well as leading to relinquishment to a shelter. Similarly to Salman et al. (1998), the most common behavioral reasons were house-soiling and "incompatibility with other pets." However, the relative proportions of these two problems were different between the studies. In this study, "incompatibility with other pets" (including "aggression between cats" and "problems with a dog") made up 8% of the "non stray" population, whereas this category accounted for 2% of the population in Salman et al. (1998). Three percent of the "non stray" population in this study displayed house-soiling, compared with 5% in the Salman et al. (1998) study. These small differences are difficult to interpret and may reflect differences in cat populations in the vicinity of shelters studied, or interpretative differences of owner responses.

Very few surveys of the general, owned cat population have been conducted, so it is not possible to make many comparisons with our data. In the USA, Morgan and Houpt (1990) surveyed 120 cat owners who had not sought any help for behavioral problems, as part of a study to investigate the effect of de-clawing on cat behavior. Of the 120 cats, 72 (60%) scratched the furniture; 51 (42%) ate houseplants; 44 (36%) showed aggression to other cats; 31 (25%) stole food; 21 (17%) hissed or were aggressive towards people; 20 (16%) house-soiled (either elimination or marking); 20 (16%) vocalized excessively; 9 (7%) chewed fabric; 5 (4%) were "shy," and 9 (8%) showed other undesirable behaviors. In an unpublished survey in the UK of 109 cat owners, 10 (9%) reported aggression to the owner, 28 (26%) reported aggression towards other cats in the household, and 14 (13%) reported house-soiling. These behaviors were also more likely to be described by owners as being a problem for them than other behavioral signs, such as scratching or avoiding contact with people (Casey, unpublished data). The proportion of cats showing aggression to other cats and house-soiling is similar to the proportion of relinquishing owners citing these behaviors, suggesting that where these behaviors occur they often lead to relinquishment, presumably because of their impact on owner lifestyle.

Because of the different ways in which "reasons for return" were categorized, it is difficult to compare all aspects of the population of cats returned to shelters in this study with those reported by Neidhart and Boyd (2002). However, the 20% of behavioral reasons reported by Neidhart and Boyd (2002) is lower than the 38% reported in this study. The proportion returned because of allergies or asthma (18% compared with 10%) is also higher in this population, although those returned for illnesses is lower (2% compared with 7%). As well as differences in local cat populations, these differences may be partly due to differences in data collection style. In this study, owners brought the cat back to the shelter where they acquired their cat, and discussed their reasons for doing so with people that they were familiar with. In Neidhart and Boyd's (2002) study, owners were called on the telephone by researchers that they were not familiar with. In addition, in Neidhart and Boyd's (2002) study, the data were collected at standard periods of time after adoption, and hence a variable time after re-relinquishment, whereas in our study, the time of data collection was at the time the cat was dropped off at the shelter. Owners may have been more willing to discuss the development or occurrence of problems with shelter staff with whom they were familiar. It is also possible that this difference reflects cultural differences in the recognition of, or tolerance for, behavioral problems in the USA and UK.

Comparing Relinquished and Returned Populations
Significant differences were found in both the age and neuter status of returned cats, compared with the remainder of the relinquished population. A difference in neuter status between the populations probably reflects the policy of shelters to neuter cats before homing, as well as being related to age differences between the populations. The age difference between the groups suggests that adult cats are more likely to be returned than kittens. This may be due to increased problems introducing adult cats into a household, especially where there are existing cats, or increased retention of kittens because of their increased adaptability to new environments (Casey and Bradshaw 2008).

There were also differences in the reasons for relinquishment in returned cats. In particular, cats appear to be more likely to be returned to shelters than relinquished in the first place, because of owner-reported behavioral problems. These differences may be due to owners feeling less guilty about, and hence more able to admit to, problems occurring in a cat that they have not had for very long. Alternatively, owners may be less worried about a good home being found for cats that they have not had time to develop a strong bond with, and hence be more willing to discuss problem behaviors. It is also possible that problem behaviors, such as nervousness, are more likely to occur in the period after a cat is first homed, as the environment is unfamiliar and unpredictable for the cat. This may particularly be the case with respect to problems occurring between the cats in the household.

Critique of Methodology
In the collection of data about "reasons for relinquishment," owners were presented with an open question, rather than a list of choices. Although the aim was to allow as much flexibility of response as possible, this did mean that some answers were difficult to categorize. It also meant that owners might have chosen to give the least controversial or embarrassing of several contributory factors leading to their decision to relinquish their cat. In addition, in some cases the relinquishment questionnaires were filled in by the shelter staff for the owner, so may have been the staff members' interpretation of the reason, rather than that which the owner actually intended. However, the large sample size should mean that the data collected provide a reasonably reliable indication of the reasons that cat owners gave to shelter staff at the time of relinquishment. The collection of data over a 12-month-period also meant that data should not have been biased by being only within, or without, the kittening season.

As mentioned in the results, the "return" data were underrepresented because information on whether cats were returns or not was missing from a large proportion of cases. Therefore the return rate was unreliable and the sample of returns for comparison with the rest of the relinquished cats was relatively small.

Significance of Findings
The data from this study provide a benchmark of primary owner reasons for relinquishment of cats in 2001 against which trends in cat relinquishment can be compared, and provide comparative baseline information for studies evaluating the effect of preventative or educative programs.

It is significant that a large number of cats entered rescue facilities in the UK at the time of the study as un-owned (i.e., stray or abandoned) individuals. Further research to identify the different origins of these cats is fundamental in informing feline welfare policies and programs of education to reduce the numbers of these cats entering facilities. Un-owned cats may be domestic animals that have been abandoned or "dumped" by their owners. However, they may also be owned cats that move away from their home environment for various reasons (e.g., to avoid areas of high cat population density), or even animals which are owned and cared for but which are brought to facilities by neighbors who are not aware of the cat's origin. They may also originate from feral or semi-feral populations occupying residential areas, or be the unwanted kittens of owned cats living peripherally to their natal home. Since different solutions are likely to be needed for cats from these different origins, further information about this category of relinquished cat is important in addressing this issue.

The significant numbers of cats relinquished because owners have too many cats, have unwanted kittens, or because their cat is pregnant emphasizes the continuing importance of neutering campaigns and clinics, and the education of owners about the importance of neutering their cats. However, further research is needed to identify the most effective targeting of resources in this area—for example, the relative effects of directing neutering campaigns in particular regions or towards owners in particular socio-economic groups, focusing neutering on feral populations, or concentrating on early neutering of kittens.

Nearly a fifth of cats enter rescue facilities because of changes in their owners' circumstances, such as financial pressures, going into rental accommodation, or illness. The significant impact on such changes on the level of relinquishment emphasizes the importance of engagement by welfare organizations in programs aimed at supporting owners to keep their pets. This might include, for example, provision of respite care whilst owners are in hospital. In addition, welfare organizations have an important role in lobbying for the provision of pet-friendly social housing and care homes for the elderly.

It is interesting to note that the proportion of cats relinquished because of asthma or allergy is not unrealistic, given the proportion of the general population reporting atopy to cat dander in other studies. This suggests that the common belief of shelter staff that allergy is an excuse given for relinquishment by owners who are unwilling to discuss potentially more embarrassing or personal reasons may only apply in a minority of cases.

Behavioral problems make up a relatively small percentage of the primary reasons why cats are relinquished to rescue shelters. However, as reported by DiGiacomo, Arluke and Patronek (1998), the occurrence of undesired behaviors may be a factor in the decision to relinquish in a greater number of cases, particularly where the behavior shown by a cat is one likely to impact on the owner's lifestyle. It should therefore also be a priority for welfare charities to have the resources to provide qualified behavioral advice for owners considering relinquishment, to increase the chance of cats being maintained in their home environment (providing this is appropriate for their welfare). The availability of suitable behavioral advice is also important, given the proportion of cats returned to shelters after homing due to the occurrence of undesirable behaviors. Staff training in the (not inconsiderable) skills needed to effectively match individual cats to the right environment may help reduce these numbers, as would the provision of specific advice to owners which is appropriate to their circumstances and the individual cat that they are homing. For example, advice on the behavioral needs of cats and possible enrichment techniques to fulfill these should be routinely provided to owners homing cats to an indoor environment. To maximize the effectiveness of such interventions, further research is needed to validate the assessment of cats and owners' attitudes and circumstances, to facilitate the matching process. Research is also needed to validate both the specific preventative advice given to owners and the format in which this is provided, to maximize compliance.

In conclusion, this study has provided some basic information about the primary reasons for relinquishment of domestic cats in the UK. The results suggest that stray or abandoned cats and unwanted kittens remain a significant problem for rescue facilities. However, change in owner circumstances, allergy or asthma, and undesirable behaviors in cats are also important reasons for relinquishment. The study has also shown that behavioral problems are an important cause of cats being returned to centers after homing, suggesting that reviewing the criteria used to match cats with owners and providing appropriate behavioral advice after homing should be priorities for cat welfare organizations.

References

Bailey, G. 1992. *Parting with a Pet Survey*. Burford, UK: Blue Cross Publication.

Casey, R. A. and Bradshaw, J. W. S. 2001. A comparison of referred feline clinical behavior cases with general population prevalence data. Paper presented at the British Small Animal Veterinary Association Congress, Birmingham, UK, 2001.

Casey, R. A. and Bradshaw, J. W. S. 2008. The effects of additional socialisation for kittens in a rescue centre on their behaviour and suitability as a pet. *Applied Animal Behaviour Science* 114(1–2): 196–205.

Clifton, M. 1993. Count finds 5 million a year—AHA says 12 million. *Animal People* October 1: 8.

DiGiacomo, N., Arluke, A. and Patronek, G. 1998. Surrendering pets to shelters: The relinquisher's perspective. *Anthrozoös* 11: 41–51.

Foucard, T. 1991. Allergy and allergy-like symptoms in 1050 medical students. *Allergy* 46: 20–26.

Luke, C. 1996. Animal shelter issues. *Journal of the American Veterinary Medical Association* 208(4): 524–527.

Miller, D. D., Staats, S. R., Partlo, C. and Rada, K. 1996. Factors associated with the decision to surrender a pet to a shelter. *Journal of the American Veterinary Medical Association* 209: 738–742.

Morgan, M. and Houpt, K. A. 1990. Feline behaviour problems: The influence of declawing. *Anthrozoös* 3: 50–53.

Neidhart, L. and Boyd, R. 2002. Companion animal adoption study. *Journal of Applied Animal Welfare Science* 5(3): 175–192.

New, J. C., Salman, M. D., King, M., Scarlett, J. M., Kass, P. H. and Hutchinson, J. M. 2000. Characteristics of shelter-relinquished animals and their owners compared with animals and their owners in U.S. pet-owning households. *Journal of Applied Animal Welfare Science* 3: 179–200.

Olson, P. N., Moulton, C., Nett, T. M. and Salman, M. D. 1991. Pet overpopulation: A challenge for companion animal veterinarians in the 1990s. *Journal of the American Veterinary Medical Association* 198: 1151–1152.

Patronek, G. J., Glickman, L. T., Beck, A. M., McCabe, G. P. and Ecker, C. 1996. Risk factors for relinquishment of cats to an animal shelter. *Journal of the American Veterinary Medical Association* 209: 582–588.

Salman, M. D., New, J. G., Scarlett, J. M., Kass, P. H., Ruch-Gallie, R. and Hetts, S. 1998. Human and animal factors related to the relinquishment of dogs and cats in 12 selected animal shelters in the United States. *Journal of Applied Animal Welfare Science* 1(3): 207–226.

Shaaban, R., Zureik, M., Soussan, D., Neukirch, C., Heinrich, J., Sunyer, J., Wjst, M., Cerveri, I., Pin, I., Bousquet, J., Jarvis, D., Burney, P. G., Neukirch, F. and Leynaert, B. 2008. Rhinitis and onset of asthma: a longitudinal population-based study. *The Lancet* 372: 1049–1057.

Young Adults' Attachment to Pet Dogs: Findings from Open-Ended Methods

Lawrence A. Kurdek[*]
Department of Psychology, Wright State University, USA

[*] *Sadly, Larry Kurdek died on June 11, 2009. An obituary can be found at: www.beyondhomophobia.com/blog/2009/06/12/remembering-larry-kurdek/*

ABSTRACT Turning to someone in times of emotional distress (safe haven) is one key feature of an attachment bond. Aspects of pet dogs as sources of safe haven were examined with open-ended methods for two samples of young adults who were college students (total $n = 566$, mean age = 19.24 years). Based on ranked nominations, relative to other features of pet dogs as attachment figures, safe haven was the least salient. Nonetheless, although participants were less likely to turn to pet dogs than to mothers, friends, and romantic partners in times of distress, they were more likely to turn to pets than to fathers and brothers and just as likely to turn to sisters. Differences between pet dogs and some humans as sources of safe haven were smallest for participants with high levels of involvement in the care of their dogs and participants who regarded their dogs as strongly meeting needs for relatedness. It is concluded that characteristics of both the dog and the owner predispose young adults to regard their dogs as a source of safe haven and serve as one basis for establishing attachment bonds with them.

Keywords: attachment, pet dogs, safe haven

Most adults establish close emotional ties to multiple people (La Guardia et al. 2000; Tancredy and Fraley 2006). Many adults also give their pet dogs a favored place in their lives in that they report strong emotional ties to them (Beck and Madresh 2008), regard them as members of the family (Albert and Bulcroft 1988), and grieve their deaths (Stallones 1994). In light of reports that the number of people who have someone to talk to about matters that are important to them has declined over the past two decades (McPherson, Smith-Lovin and Brashears 2006), factors that buffer the negative effects of social isolation—such as a strong attachment to a pet dog—are of potential practical importance. In the current study, I addressed aspects of the extent to which young adults establish attachment bonds with their pet dogs.

I selected young adults in the age group of 18 to 25 years for substantive reasons. In the course of negotiating developmental tasks regarding individuation and intimacy, young adults typically reorganize their attachment systems by decreasing ties to parents and siblings and increasing ties to friends and romantic partners (Trinke and Bartholomew 1997). Because this reorganization may heighten the need for autonomy, young adults may perceive parents and siblings as being critical and judgmental, raising the possibility that they value pet dogs for their unconditional affection (Archer 1997). I selected dogs over other pets because there are more than 60 million pet dogs in the United States alone (American Veterinary Medical Association 2002), and perceived support from them tends to be stronger than that from other pets (Bonas, McNichols and Collis 2000).

I grounded the concept of attachment bonds in Ainsworth's (1991) model of attachment features. Although originally developed to study parent–infant relationships, this model also has been used to study adult relationships (e.g., Tancredy and Fraley 2006). According to this model, attachment figures have four features. These include being physically near and accessible (*proximity maintenance*), being missed when absent (*separation distress*), being a dependable source of comfort (*secure base*), and being sought for contact and assurance in times of emotional distress (*safe haven*). These features are relevant to a distinction between caregiving bonds and attachment bonds (George and Solomon 1996).

Caregiving bonds highlight the features of proximity maintenance and separation distress and focus on providing sensitive and responsive care. Based on evidence that owners both enjoy the companionship of their pet dogs (Bonas, McNichols and Collis 2000; Kurdek 2008) and miss their dogs when away from them (Archer and Winchester 1994; Kurdek 2008), it is reasonable to conclude that owners form caregiving bonds with their pet dogs. In contrast, attachment bonds highlight the features of secure base and especially safe haven and focus on using the attachment figure to regulate threats to felt security (Simpson and Rholes 2000). Owners regard pet dogs as being highly reliable and dependable (Bonas, McNichols and Collis 2000; Kurdek 2008), but there is limited evidence that they actually turn to their dogs for support in times of emotional distress (Flynn 2000; Beck and Madresh 2008).

In the current study, I addressed this limitation by examining the extent to which young adults report that their pet dogs show the feature of safe haven. In contrast to previous studies of attachment to pets that have used structured ratings (e.g., Beck and Madresh 2008; Kurdek 2008), in the present study, I used open-ended methods in which participants freely nominated and prioritized people or pets who met criteria for specific attachment features. Such methods are often used to study how participants themselves structure their close relationships (e.g., Doherty and Feeney 2004).

The current study had three purposes and involved two independent samples. The first purpose (using the first sample) was to compare ranked nominations regarding the four features of attachment for pet dogs. Consistent with the view that pet dogs have inherent cognitive, verbal, and behavioral limitations in actively removing sources of distress (Archer 1997) and previous findings from ratings (Kurdek 2008), I expected that safe haven would be the least prominent feature of pet dogs as attachment figures.

The second purpose of this study (using the second sample) was to compare rankings of safe haven for pet dogs to those of six human attachment figures (mothers, fathers, brothers, sisters, friends, and romantic partners). My interest here was in how pet dogs fared as sources of safe haven relative to human members that typically populate the attachment hierarchies of young adults (Trinke and Bartholomew 1997). Previous studies using rankings have shown that mothers, friends, and romantic partners are more likely than fathers and siblings to be sources of safe haven (Trinke and Bartholomew 1997). Thus, I expected that pet dogs would be less likely to be regarded as sources of safe haven than mothers, friends, and romantic partners were. However, how pet dogs would fare relative to fathers and siblings was not clear.

In line with evidence that owners vary in the strength of their attachment to their pet dogs (Archer 1997; Kurdek 2008), the final purpose of this study (also using the second sample) was to account for some of this variability by exploring what variables accentuated a preference for pet dogs over human figures as sources of safe haven. I selected two such moderator variables: *Involvement in Care of the Pet Dog* and *Fulfillment of Relatedness Needs*. Involvement in Care of the Pet Dog was of interest because caregiving bonds and attachment bonds tend to be related (Carnelley, Pietromonaco and Jaffe 1996) and because primary caregivers report more intimacy with their pets than do non-primary caregivers (Holcomb, Williams and Richards 1985) as well as stronger attachment to them (Kurdek 2008). Fulfillment of Relatedness Needs was of significance because attachments tend to be strong toward responsive human and canine partners who satisfy the need to be cared for and connected (La Guardia et al. 2000; Kurdek 2008). In sum, pet dogs might be preferred as sources of safe haven over some humans for those who have high levels of involvement in the care of their dog and those who regard their pet dog as strongly satisfying needs pertaining to relatedness.

Methods
Participants
Sample 1: Participants were 401 undergraduate students (289 females). Students from this sample and the second sample were enrolled in an Introductory Psychology course, received partial course credit for participation, and lived full-time with a pet dog that was at least one year old. Students from sample 1 nominated and ranked persons or pets for each of the four attachment features. Because no information was obtained on which possible human attachment figures were actually available, only data for pet dogs were of interest. Students' mean age was 19.16 years ($SD = 1.70$), and 75% were White. Of the 401 dogs rated, 65% were identified as pure breeds (with the most frequent breed being Labrador Retriever, 11%); 49% were female; and 69% were spayed or neutered. The mean age of the dogs was 5.33 years ($SD = 3.78$).

Sample 2: Participants were 165 students (112 females) who nominated and ranked persons or pets only as sources of safe haven. They also rated the extent to which they took care of their dogs and the extent to which their dogs met needs regarding relatedness. Nominations and rankings were performed before ratings, so the latter did not influence the former. Students' mean age was 19.32 years ($SD = 1.85$), and 78% were White. The number of students who had mothers, fathers, brothers, sisters, friends, and romantic partners available was 163, 154, 110, 107, 159, and 112, respectively. Of the 165 dogs rated, 58% were identified as pure breeds (with the most frequent breed being Labrador retriever, 11%); 49% were female; and 69% were spayed or neutered. The mean age of the dogs was 5.42 years ($SD = 3.62$).

Measure of Demographic Variables
Participants from each sample provided information about their age, gender, and race. They also described the age, sex, and breed of the dog they evaluated and indicated whether this dog was spayed or neutered. All measures were completed in small groups on a university

campus and were administered by an undergraduate research assistant. Participants were told that they needed to be at least 18 years of age, that the survey would take about 30 minutes to complete, that their responses were confidential, that they would complete the surveys anonymously, and that the completion of the survey implied informed consent.

Ranking Measures

Two ranking measures were used, one focusing on all four features of attachment and the other only on the feature of safe haven.

All Attachment Features: For each of eight questions, participants from sample 1 were instructed to list the first name of up to five individual persons or pets. They were told that they could list their dog as well as any other pet and that they did not have to use all five lines provided. The first name listed indicated the first choice, the second name indicated the second choice, and so on. Next to each name, participants identified the relationship the listed entry had with them (e.g., father, dog).

The eight questions were developed by Doherty and Feeney (2004), included two questions for each feature of attachment, and were presented in the order indicated here. For *safe haven*, the questions were "Who do you talk to when you are worried about something or when something bad happens to you?" and "Who do you turn to for comfort when you are feeling upset or down?" For *secure base*, the questions were "Who do you feel will always be there for you, if you needed them?" and "Who do you feel you can always count on, no matter what?" For *proximity maintenance*, the questions were "Who do you like to spend time with?" and "Who is it important for you to see/talk with regularly?" Finally, for *separation distress*, the questions were "Who do you not like to be away from?" and "Who do you miss the most during separations?"

I coded the identified relationship for each listed name as one of seven attachment figures (mother, father, brother, sister, friend, romantic partner, and dog). Because more than one brother, sister, friend, and dog could be named, individuals sharing the same relationship status were given unique codes. Preliminary analyses indicated that the first-named brother, sister, friend, and dog was most frequently used throughout the survey, so I used only these entries to derive scores.

I weighted the order in which names were listed with a value of 5 through 1 for the first through last rankings, respectively. Relationships not listed with a 0 were weighted. Scores were derived for each feature for pet dogs by averaging the weighted rankings assigned to dogs for the two questions pertaining to each feature. These scores ranged from 0 through 5. Here and later, high scores reflected high levels of the construct measured. The reliability of the summed composite scores for pet dogs, as assessed by Cronbach's α, was 0.62 for safe haven, 0.73 for secure base, 0.62 for proximity maintenance, and 0.87 for separation distress.

Safe Haven Feature: I adapted the second assessment of safe haven from Ryan et al.'s (2005) measure of emotional reliance. A pilot sample of 121 college students (different from the other samples used in this study) rated dogs, mothers, fathers, brothers, sisters, best friends, and romantic partners on Ryan et al.'s entire 10-item Emotional Reliance Scale, in order to identify a smaller set of meaningful items. Respondents rated the extent to which they would turn to target person or animal when experiencing two positive emotions (e.g., feeling happy or having good news) and eight negative emotions (e.g., feeling alone or depressed).

Principal components analyses revealed one-factor solutions for dogs, brothers, sisters, and best friends and two-factor solutions for mothers, fathers, and partners. For each two-factor solution, the two positive emotions defined one factor, and the eight negative emotions defined the other factor. Because safe haven concerns seeking comfort in times of distress, I selected the four items that had the highest loadings on the negative emotions factor across all figures. Also, because the focus of this study was on actual rather than possible sources of comfort, all items were worded as "Who *do* you turn to" rather than "Who *would* you turn to." The items were: (a) Who do you turn to when you are lonely or depressed? (b) Who do you turn to when you are feeling very bad about yourself and need a boost? (c) Who do you turn to when you are feeling overwhelmed by responsibilities and commitments? and (d) Who do you turn to when you are disappointed?

For each question, participants from sample 2 were instructed to list the first name of up to five individual persons or pets. They were told that they could list their dog as well as any other pet and that they did not have to use all five lines provided. The first name listed indicated the first choice, the second name indicated the second choice, and so on. Next to each name, participants identified the relationship the listed individual had with them. I weighted the order in which names were listed with a value of 5 through 1 for the first through last listings, respectively. Relationships not listed with a 0 were weighted. I obtained scores that ranged from 0 through 5 for dogs and each available human figure by averaging the weighted rank for each figure. Cronbach's α was 0.73 for pet dogs, 0.75 for mothers, 0.65 for fathers, 0.75 for brothers, 0.72 for sisters, 0.75 for friends, and 0.83 for partners.

Rating Measures

Involvement in Care of the Pet Dog: Participants from sample 2 completed Kurdek's (2008) measure of dog care by indicating how much they agreed (1 = strongly disagree, 5 = strongly agree) with each of 10 statements (e.g., I am the one most likely to feed my dog); Cronbach's α was 0.93.

Fulfillment of Relatedness Needs: Participants from sample 2 indicated how true it was (1 = not at all true, 7 = very true) that their pet dog met needs for relatedness. The three items were derived from La Guardia et al.'s (2000) Need Satisfaction Scale (e.g., When I am with my dog, I feel loved and cared about); Cronbach's α was 0.70.

Results

The Most Salient Feature of Pet Dogs as Attachment Figures (Sample 1)

The first purpose of this study was to assess the salience of safe haven as a feature of pet dogs relative to the other three features of an attachment figure. To provide a qualitative sense of how dogs were ranked across all features, I present the percentage of respondents ranking dogs first, second, third, fourth, or fifth by type of feature in Table 1. Consistent with these percentages, the means (and *SD*s) for the weighted ranked features, from highest to lowest, were 0.55 (0.68) for separation distress, 0.35 (0.42) for proximity maintenance, 0.34 (0.44) for secure base, and 0.26 (0.33) for safe haven. A one-way (feature) multivariate analysis of variance (MANOVA) indicated that the four mean rankings were not equivalent ($F_{(3, 389)}$ = 28.36, $p < 0.01$). As expected, post-hoc comparisons using the Bonferroni correction ($p < 0.05$) indicated that rankings for safe haven were lower than those for each of the other three features. In addition, rankings for separation distress were higher than those for each of the other three features. Rankings for secure base and proximity maintenance were equivalent.

Table 1. Percentage of respondents from sample 1 (*n* = 410) ranking dogs first, second, third, fourth, or fifth for each feature of attachment

Feature	Ranking				
	1	2	3	4	5
Safe Haven	3.8	7.5	10.0	10.5	6.8
Secure Base	7.0	8.4	9.8	10.6	9.5
Proximity Maintenance	5.0	9.6	12.2	12.7	10.9
Separation Distress	7.8	18.4	13.9	11.3	10.3

Differences in Rankings of Human Figures Versus Dogs on Safe Haven (Sample 2)

The second purpose of this study was to assess the likelihood that pet dogs, rather than each of six human attachment figures, were turned to for safe haven. To provide a qualitative sense of how dogs were ranked relative to human figures, I present the percentage of respondents ranking dogs and each human figure first, second, third, fourth, or fifth in Table 2. Consistent with these percentages, from highest to lowest, relevant means (and *SE*s) were 0.81 (0.04) for mothers, 0.78 (0.06) for partners, 0.61 (0.03) for friends, 0.41 (0.03) for dogs, 0.35 (0.03) for sisters, 0.28 (0.02) for fathers, and 0.14 (0.02) for brothers.

Table 2. Percentage of respondents from sample 2 ranking each figure first, second, third, fourth, or fifth for safe haven.

Figure	Ranking					*n*
	1	2	3	4	5	
Dog	7.2	12.3	9.9	8.7	6.5	165
Mother	31.7	15.2	9.7	4.6	1.5	163
Father	5.7	10.1	5.5	4.3	3.2	154
Brother	2.0	5.7	4.8	3.2	2.0	110
Sister	4.7	12.2	10.1	7.3	3.5	107
Friend	17.6	34.1	10.0	4.7	2.0	159
Partner	40.4	16.6	18.3	2.5	1.4	112

Note: *n* indicates the number of respondents who had the figure available to them.

I conducted a two-level hierarchical linear regression in which attachment figures (level 1) were nested in participants (level 2). I decomposed the main effect for figure into a set of six simple contrasts in which mothers, fathers, brothers, sisters, best friends, and romantic partners were each compared with pet dogs. A negative coefficient for each contrast indicated that the score for dogs was higher than that for the target human figure, whereas a positive coefficient indicated that the score for dogs was lower than that for the target human figure. Dummy variables for each of the seven figures were treated as random effects, resulting in an unconditional error covariance matrix. Evidence of significant variability for a random effect would provide a statistical basis for accounting for this variability with moderator variables. I treated each simple contrast as a fixed effect.

Because the main effect for figure involved multiple contrasts, chance significant findings were controlled with an initial multivariate test in which all contrasts were tested simultaneously. This multivariate test was significant ($\chi^2_{(6)}$ = 352.77, *n* = 165, *p* < 0.01).

Follow-up univariate tests revealed that only the simple contrast involving sisters was not significant, indicating that pet dogs were ranked equivalently to them. As can be derived from the means for each figure, with regard to the other simple contrasts, whereas dogs were ranked less highly than mothers ($\beta = 0.39$), friends ($\beta = 0.20$), and partners ($\beta = 0.37$), they were ranked more highly than fathers ($\beta = -0.12$) and brothers ($\beta = -0.27$) ($ps < 0.01$). Moreover, the random effects for each contrast were significant (all $ps < 0.01$), indicating that there was significant variability to be accounted for by moderator variables.

Moderators of Differences Between Human Figures and Pet Dogs on Safe Haven (Sample 2)

The final purpose of this study was to explore whether differences in the extent to which pet dogs versus each of six human attachment figures were turned to during distress varied by two moderator variables. I accomplished this with two analyses (one for each moderator variable) in which I added two sets of predictors to the two-level hierarchical linear regression described in the previous section. The first set included the mean-centered predictor representing the main effect for the moderator variable. The second set included predictors representing the interaction between each simple contrast and the mean-centered moderator variable. All interaction terms were derived by multiplying relevant predictors.

The analysis of moderator effects occurred in three steps. First, because each moderator effect involved multiple terms, chance significant findings were controlled by first conducting a multivariate test in which all contrasts relevant to the interaction were tested simultaneously. Second, if the multivariate interaction effect was significant, I then conducted univariate tests to identify the specific simple contrasts for which interactions occurred. Finally, if the interaction term involving a simple contrast was significant, I performed tests of simple slopes in which I derived the unstandardized coefficient that represented the difference between the targeted human figure and dogs at three levels of the moderator variable. These were low levels (one standard deviation below the mean), average levels (at the mean), and high levels (one standard deviation above the mean). I present findings for each step for each moderator variable.

Involvement in Care of Pet Dog: The multivariate interaction test was significant ($\chi^2_{(6)} = 24.16$, $n = 165$, $p < 0.01$) and remained so even with controls for effects attributed to relatedness needs ($\chi^2_{(6)} = 16.22$, $p < 0.01$). Univariate tests showed that interactions were significant for five of the six contrasts: mothers versus dogs, $\beta = -0.30$; fathers versus dogs, $\beta = -0.19$; brothers versus dogs, $\beta = -0.19$; sisters versus dogs, $\beta = -0.13$; and friends versus dogs, $\beta = -0.20$ ($ps < 0.05$). The simple slopes for these five contrasts are shown at the top of Table 3. At high levels of involvement, preferences for mothers over dogs were less striking than at low and average levels; preferences for dogs over fathers, brothers, and sisters were more striking at high levels than at either low or average levels; and, at high levels, dogs were preferred equally to friends.

Fulfillment of Relatedness Needs: The multivariate interaction test was significant ($\chi^2_{(6)} = 25.27$, $n = 165$, $p < 0.01$) and remained so even with controls for effects attributed to involvement in care ($\chi^2_{(6)} = 15.27$, $n = 165$, $p < 0.05$). Univariate tests showed that interactions were significant for each of the six contrasts: mothers versus dogs, $\beta = -0.11$; fathers versus dogs, $\beta = -0.13$; brothers versus dogs, $\beta = -0.15$; sisters versus dogs, $\beta = -0.10$; friends versus dogs, β = –0.13, and partners versus dogs, β = –0.24 ($ps < 0.05$). The simple slopes for each contrast are shown at the bottom of Table 3. At high levels of need satisfaction, preferences for mothers over dogs were less striking than at low and average levels; at high levels, preferences for dogs over fathers, brothers, and sisters were more striking than at either low or average levels; and, at high levels, dogs were preferred equally to both friends and partners.

Table 3. Unstandardized coefficients for simple slopes involving contrasts between human figures and pet dogs on rankings of safe haven for sample 2.

Human Figure	Low	Average	High
		Involvement in Care of Dog	
Mother	0.63**	0.38**	0.14*
Father	0.02	–0.13**	–0.28**
Brother	–0.11*	–0.27**	–0.42**
Sister	0.04	–0.07	–0.17**
Friend	0.36**	0.20**	0.03
		Relatedness Needs	
Mother	0.51**	0.39**	0.26**
Father	0.01	–0.13**	–0.27**
Brother	–0.09	–0.27**	–0.44**
Sister	0.05	–0.07	–0.18**
Friend	0.34**	0.20**	0.05
Partner	0.64**	0.37**	0.11

*$p < 0.05$, **$p < 0.01$
Note: Positive values indicate that the ranking for dogs was lower than that for the human figure. Negative values indicate that the ranking for dogs was higher than that for the human figure.

Discussion

The Most Salient Feature of Pet Dogs as Attachment Figures

Consistent with previous studies of human attachment figures (Trinke and Bartholomew 1997; Doherty and Feeney 2004; Kurdek 2008), pet dogs were not nominated equally often for the four features of attachment. As predicted, and consistent with the view that pet dogs have inherent cognitive, verbal, and behavioral limitations in actively removing sources of distress (Archer 1997), safe haven was the least prominent feature of pet dogs as attachment figures. Because Kurdek (2008) also found that safe haven was the least prominent feature for college students based on ratings, this finding appears to be a robust one, at least for college students. In line with evidence that pet owners miss their dogs when separated from them and grieve their deaths (Archer and Winchester 1994; Kurdek 2008), separation distress was the most salient feature.

Differences in Rankings of Safe Haven Between Human Figures and Dogs

The likelihood that pet dogs were turned to rather than human figures in times of emotional distress depended on the type of human figure. Dogs were less likely to be turned to than mothers, friends, and partners; as likely to be turned to as sisters; and more likely to be turned to fathers and brothers. These findings are consistent with those of other studies using rankings or ratings (Trinke and Bartholomew 1997; Doherty and Feeney 2004; Tancredy and Fraley

2006) in showing that whereas romantic partners, best friends, and mothers are widely regarded by young adults as prominent attachment figures, fathers and siblings are not.

The current findings also extend the literature by showing that although mothers in particular are a preferred source of safe haven relative to pet dogs, within families, dogs fare as well or better than fathers, sisters, and brothers. From this standpoint, dogs, indeed, function as family members (Albert and Bulcroft 1988). Given current clinical interest in the factors that mitigate the negative effects of social isolation (McPherson, Smith-Lovin and Brashears 2006), future work could examine whether strong attachments to pet dogs buffer any of the negative effects associated with loneliness and depression across the adult lifespan and not just for the elderly (Enders-Slegers 2000).

Future work is needed to address additional aspects of pet dogs as sources of safe haven. Open-ended surveys and daily diary methods could be used to obtain samples of the kinds of instances in which dogs are turned to, and owners also could evaluate how effective the dog was in mitigating emotional distress. Because dogs have inherent limitations in their ability to manage causes of distress, information could also be obtained on what specific features of the dog provide comfort as well as why available human figures either were not sought or were found to be wanting. It is possible that the unconditional nature of the support that pet dogs provide is key to dogs being regarded as a preferred source of emotional comfort.

Moderators of Differences in Rankings of Safe Haven Between Human Figures Versus Dogs

Consistent with Kurdek's (2008) findings based on ratings, differences in rankings between human figures and pet dogs on safe haven varied by characteristics of both the owner and the dog. With regard to owners, those who were highly involved in the care of their dog were likely to prefer pet dogs over fathers, brothers, and sisters, and were as likely to prefer dogs as they did friends. They were less likely to prefer mothers over dogs than those with lower levels of involvement. Only partners outranked dogs at all levels of involvement.

It is possible that high levels of involvement in routine caregiving tasks provide opportunities for attachment bonds to develop much as sensitive and responsive caregiving provides a context for secure attachments to develop for dyads involving parents and infants (Ainsworth 1991). Longitudinal work would be useful in assessing whether attachment bonds later provide one motivation for sustaining those caregiving activities as pet dogs inevitably grow old, become infirm, and require fairly intensive and costly care.

With regard to characteristics of the pet dog, owners who regarded their dogs as strongly meeting needs for relatedness were likely to prefer pet dogs over fathers, brothers, and sisters, and were as likely to prefer dogs as they did both friends and partners. They were less likely to prefer mothers over dogs than those who did not regard their dogs as meeting needs for relatedness. This finding is of note because ratings of relatedness have been predictive of relational outcomes involving both humans (Ryan et al. 2005) and pet dogs (Kurdek 2008) independent of other needs such as autonomy and competence. Perhaps dogs fairly easily satisfy relatedness needs because they are regarded as uncritical and unconditional providers of affection and acceptance.

Taken together, the findings regarding moderation indicate that characteristics of both the owner and the pet dog independently influence whether a pet dog is regarded as a source of safe haven. Future studies could identify additional moderator variables and develop conceptual models for integrating their joint influences. It is possible, for example, that the need for relatedness is the most proximal link to pet dogs as sources of safe haven and that other variables influence dogs as sources of safe haven through relatedness needs. Path models with longitudinal data that track the development of owner–pet dog interactions would be useful in validating such meditational models.

Limitations and Conclusion

The current study has at least four limitations. First, because all participants were college students, it is unclear whether the findings obtained here will generalize to samples that are more diverse with regard to both developmental status and educational background. Data from community samples are especially needed. Second, because all data came from participants, responses could have been influenced by biases common to self-report measures. Third, although rankings provided an opportunity for participants to describe the structure of their attachment hierarchies in an open-ended manner, they lacked a standard metric that is needed to interpret differences in ranks between figures. Finally, although the reciprocal nature of bonds between pet owners and their dogs is acknowledged, the attachment of pet dogs on human owners was not studied.

Despite these limitations, the current study advances the study of human–pet interactions within an attachment context in at least two ways. First, it indicates that the ranking method can be easily modified to include pets as possible attachment figures. Second, it indicates that although safe haven is not the most prominent feature of pet dogs as attachment figures, individual difference variables targeting both the owner and the dog affect the extent to which young adults prefer their dog over some human figures as a source of safe haven and establish attachment bonds with them.

Acknowledgements

Appreciation is extended to Patrick Przyborowski, Jennifer Dazey, Kristi Lytton, and Lisa Brown for their assistance in data collection, data coding, and data entry.

References

Ainsworth, M. D. S. 1991. Attachments and other affectional bonds across the life cycle. In *Attachment Across the Lifecycle,* 33–51, ed. C. M. Parkes, J. Stevenson-Hinde and P. Marris. New York: Routledge.

Albert, A. and Bulcroft, K. 1988. Pets, families, and the life course. *Journal of Marriage and the Family* 50: 543–552.

American Veterinary Medical Association. 2002. *U.S. Pet Ownership & Demographics Sourcebook.* Schaumburg, IL: Author.

Archer, J. 1997. Why do people love their pets? *Evolution and Human Behavior* 18: 237–259.

Archer, J. and Winchester, G. 1994. Bereavement following death of a pet. *British Journal of Psychology* 85: 259–271.

Beck, L. and Madresh, E. A. 2008. Romantic partners and four-legged friends: An extension of attachment theory to relationships with pets. *Anthrozoös* 21: 43–56.

Bonas, S., McNichols, J. and Collis, G. M. 2000. Pets in the networks of family relationships: An empirical study. In *Companion Animals and Us: Exploring the Relationships Between People and Pets*, 209–236, ed. A. L. Podberscek, E. S. Paul and J. A. Serpell. Cambridge: Cambridge University Press.

Carnelley, K. B., Pietromonaco, P. R. and Jaffe, K. 1996. Attachment, caregiving, and relationship functioning in couples: Effects of self and partner. *Personal Relationships* 3: 257–278.

Doherty, N. A. and Feeney, J. A. 2004. The composition of attachment networks throughout the adult years. *Personal Relationships* 11: 469–488.

Enders-Slegers, M. 2000. The meaning of companion animals: Qualitative analysis if the life histories of elderly cat and dog owners. In *Companion Animals and Us: Exploring the Relationships Between People and Pets*, 237–256, ed. A. L. Podberscek, E. S. Paul and J. A. Serpell. Cambridge: Cambridge University Press.

Flynn, C. P. 2000. Woman's best friend: Pet abuse and the role of companion animals in the lives of battered women. *Violence Against Women* 6: 162–177.

George, C. and Solomon, J. 1996. Representational models of relationships: Links between caregiving and attachment. *Infant Mental Health Journal* 17: 198–216.

Holcomb, R., Williams, R. C. and Richards, P. S. 1985. The elements of attachment: relationship maintenance and intimacy. *Journal of the Delta Society* 2: 28–33.

Kurdek, L. A. 2008. Pet dogs as attachment figures. *Journal of Social and Personal Relationships* 25: 247–266.

La Guardia, J. G., Ryan, R. M., Couchman, C. E. and Deci, E. L. 2000. Within-person variation in security of attachment: A self-determination theory perspective on attachment, need fulfillment, and well-being. *Journal of Personality and Social Psychology* 79: 367–384.

McPherson, M., Smith-Lovin, L. and Brashears, M. E. 2006. Social isolation in America: Changes in core discussion networks over two decades. *American Sociological Review* 71: 353–375.

Ryan, R. M., La Guardia, J. G., Solky-Butzel, J., Chirkov, V. and Kim, Y. 2005. On the interpersonal regulation of emotions: Emotional reliance across gender, relationships, and cultures. *Personal Relationships* 12: 145–163.

Simpson, J. A. and Rholes, W. S. 2000. Caregiving, attachment theory, and the connection theoretical orientation. *Psychological Inquiry* 11: 114–117.

Stallones, L. 1994. Pet loss and mental health. *Anthrozoös* 7: 43–54.

Tancredy, C. M. and Fraley, R. C. 2006. The nature of adult twin relationships: An attachment-theoretical perspective. *Journal of Personality and Social Psychology* 90: 78–93.

Trinke, S. J. and Bartholomew, K. 1997. Hierarchies of attachment relationships in young adulthood. *Journal of Social and Personal Relationships* 14: 603–625.

Angel, Rescued by ASPCA Humane Law Enforcement 5/22/08

DIALING 911 IS NOT AN OPTION FOR ANIMALS.

Answer the call to help stop animal cruelty.

Unfortunately, abused animals can't pick up a phone and call for help. That's why we at the American Society for the Prevention of Cruelty to Animals® have been leading the fight against animal abuse all across the nation, every day since 1866. But we can't do it alone. We depend on your support to create real change and give abused animals the help, hope and healing they deserve.

Learn more and join the fight against animal cruelty at aspca.org/endcruelty

ASPCA®
WE ARE THEIR VOICE.®

© 2009 ASPCA®. All rights reserved.

Empathic Differences in Adults as a Function of Childhood and Adult Pet Ownership and Pet Type

Beth Daly and L. L. Morton
Faculty of Education, University of Windsor, Ontario, Canada

Address for correspondence:
Beth Daly,
Faculty of Education,
University of Windsor,
401 Sunset Avenue,
Windsor, Ontario, N9B 3P4,
Canada.
E-mail:
bethdaly@uwindsor.ca

ABSTRACT Prompted by interesting but ambiguous findings that empathic differences in children may relate to pet preference and ownership, we extended the issue to an adult population. We investigated empathic-type responses in adults who lived with cats and/or dogs in childhood (Child-Pet) and currently (Adult-Pet), using the Interpersonal Reactivity Index (IRI), the Empathy Quotient (EQ), and the Animal Attitude Scale (AAS). Multivariate analyses of covariance, with Sex as the covariate (MANCOVA), revealed differences on the AAS, the IRI-Personal Distress scale, and the EQ-Social Skills factor. For the Child-Pet data, the Dog-Only and the Both (dog and cat) groups, compared with those in the Neither (no dog or cat) group, scored lower on the IRI-Personal Distress scale and higher on the EQ-Social Skills factor. On the AAS, all three pet groups (Dog-Only, Cat-Only, and Both) had higher ratings than the Neither group. For Adult-Pet data, the analyses revealed the Dog-Only group was lower on Personal Distress than the Neither group, and higher on Social Skills than the Neither group and the Cat-Only group. On the AAS, the Neither group was lower than all three pet-owning groups, like the childhood data, but strikingly, adults with both dogs and cats were higher on the AAS. The findings support research linking companion animals with empathic development. They warrant the continued exploration of the nature of empathic development (i.e., nature vs. nurture) and contribute to the increasing research field exploring the value of companion animals.

Keywords: adult empathy, cat–dog differences, companion animals, empathic differences, pet ownership

❖ While investigations into the emotional and psychological benefits of pets initially emerged as a largely intuitive and observed phenomenon (Levinson 1978), growing interest in this area has also led to an increase in empirical data related to various effects related to psychological, emotional, and physical well-being. Recent empirical research in the area of human–animal interactions is expanding the growing field by pointing to some positive physical and emotional benefits that pets provide for individuals. Aside from the commonsense notion that owners of dogs and cats would likely be more active in caring for their pets (e.g., feeding, grooming, walking, playing with, and so on), researchers have repeatedly pointed to empirical support for increased physical health benefits, such as cardiovascular improvement (Friedmann, Locker and Lockwood 1993; Friedmann and Thomas 1995; Friedmann et al. 2007).

One specific psychological area of investigation has focused on the development and presence of empathy in both child and adult populations. An early study (Melson, Peet and Sparks 1992) explored the relationship between children's attachment level to their pets and the development of empathy, which may well have provided the impetus for further investigation into the nature of the child–pet relationship (e.g., Melson 2003). Others have also sought to explore, from various perspectives, the nature of the child–pet relationship.

In recent years, several school-based studies have pointed to the positive role that the presence of an animal plays in a classroom setting. For instance, some studies have examined the use of animals in humane education. Although live animals were not employed, Ascione (1992) reported that positive attitudes toward animals were increased in fourth- and fifth-graders as a result of a one-year humane education program taught in classrooms. A follow-up study using Year 1 and Year 2 (Ascione and Weber 1996) post-tests suggested that human-directed empathy was correlated with some aspects of the quality of a child's relationship with animals. Another school-based intervention pointed to enhanced empathy in school children as related to having pets in the classroom (Hergovich et al. 2002). The presence of a dog in the classroom for a three-month period was correlated with enhanced empathy, compared with a control classroom that did not have a visiting dog.

Actual research involving the presence of animals in the classroom is limited, though studies examining empathy in children as a function of their relationships with animals—often, their own pets—have increasingly emerged in the past decade. Some research suggests that it is common for parents to obtain pets for their children because they feel that the presence of animals positively contributes to their children's emotional development, including empathic behaviors (Fifield and Forsythe 1999; Grier 1999). Indeed, early researchers have pointed to positive emotional development as a result of pet-keeping in childhood (Poresky and Hendrix 1990; Vizek-Vidović, Štetić and Bratko 1999).

While much research investigating empathy as it relates to animals focuses on children as the research subjects, there are also many studies which focus on empathy in adults as a function of relationships with, and attitudes toward, animals, both in their adult lives and retrospectively. Inherent in empathy research is also the question of whether human- and animal-related empathy are linked. In an early study, Paul and Serpell (1993) reported that individuals who had kept pets in childhood, compared with those who hadn't, had more humane attitudes toward animals as adults and more humane attitudes toward other people. In a later study, Paul (2000a) found that human- and animal-directed empathy were linked, though, not surprisingly, empathy-inducing sources were correlated with higher scores. For instance, pet ownership positively impacted on animal-related empathy, whereas individuals with children were higher in human-directed empathy. The nature of human- and animal-directed empathy was discussed in an extensive review by Paul (2000b), pointing to the problem with consistency in anthrozoological research. This was attributed to varying factors, such as individuals' attraction to certain species or breeds, regarding pets as a status symbol or "surrogate child," childhood history with pets, and even gender, as females have been found to be generally more empathic than males. As such, it was suggested that reported correlations between animal- and human-directed empathy should be interpreted cautiously, with due attention given to social and emotional factors that may contribute to a predisposition toward one or the other.

More recently, Taylor and Signal (2005) reported a significant relationship between human–human empathy and attitudes toward animals, specifically, that individuals who demonstrated higher empathic concern exhibited more welfare-related attitudes toward animals. The results suggest that higher human-directed empathy does correlate with more positive attitudes toward animals. Further investigating the relationship between human-directed empathy and animal attitudes, Signal and Taylor (2007) examined these variables from both a general community sample and an animal protection community sample, and again found a modest correlation between human-directed empathy and attitudes toward the treatment of animals. Distinguishing between two empathy subscales related to empathic concern and perspective-taking, it was also found that the relationship between empathic concern and the treatment of animals was higher in the animal protection community. Two explanations were offered: first, that this group adheres strongly to a specific philosophy with inflexible shifts in viewpoints; and second, that perhaps they share an elevated concern for both human and non-human animals.

Recent research has extended animal- and human-directed empathy as it relates to experiences with animal abuse. Henry (2006) found no differences on empathy scales between animal-abusers and non-abusers, which led to the question of whether human-directed empathy relates to animal abuse. In a previous study, Henry (2004) found only limited support for a relationship between individuals who had been exposed to animal cruelty and a general concern for living things. However, in a recent study, Daly and Morton (2008) found differences on the perspective-taking (PT) and personal-distress (PD) subscales of empathy of the Interpersonal Reactivity Index (IRI) (Davis 1980) for individuals who had witnessed multiple experiences of animal abuse: they were lower in personal distress scores (the extent to which observing another's negative experiences causes one's own anxiety and discomfort) yet higher on perspective-taking, which measures the ability to adopt another person's viewpoint. A "dissociation hypothesis" was put forward, suggesting that observing multiple acts of animal abuse mediates between the cognitive PT domain, and the affective PD domain. Viewing these results in a positive light, it is possible that those who have witnessed multiple acts of abuse experience a heightened disposition to cognitive empathy which assists in regulating (depressing) the emotional component, a relationship that would certainly be beneficial to those working in animal welfare environments, and perhaps even in helping professions related to people.

Pets do seem to impact on humans as children and as adults. For example, Daly and Morton (2006) reported that children with positive attitudes towards animals exhibited higher empathy than those with negative attitudes toward animals. Previous findings have pointed to empathic differences as a function of children's preference for a dog versus a cat (Daly and Morton 2003). The dog–cat differences are intriguing, and raise the question concerning whether or not these differences persist in adulthood. Thus, the purpose of the present research was to investigate whether there are empathic differences in adults who had pets in childhood as opposed to those who did not, and whether there are empathic differences in adults who currently have pets as opposed to those who do not.

Methods

Instruments

Animal Attitude Scale (AAS) (Herzog, Betchart and Pittman 1991): The Animal Attitude Scale (AAS) is a 20-item measure of attitudes toward animals; responses are given on a 5-point Likert scale, ranging from "Strongly Agree" to "Strongly Disagree." It is reported to have high internal consistency (Signal and Taylor 2007).

Empathy Quotient (EQ) (Lawrence et al. 2004): The Empathy Quotient (EQ) is a self-report measure of empathy originally developed by Baron-Cohen and Wheelright (2004) as a clinical application sensitive to a lack of empathy in individuals with a psychopathology, most notably individuals with reported autism spectrum disorders. However, Lawrence et al. (2004) conducted a series of studies confirming the reliability and validity of the scale in healthy populations as well, reporting a three-factor structure which was deemed "a better fit" in a psychometric analysis of the scale (Muncer and Ling 2006). The three factors of empathy are: (1) cognitive empathy, which examines the appreciation of affective states; (2) emotional reactivity, which is the tendency toward emotional reactions as a response to other individuals' mental states; and (3) social skills, which measures the use of skills related to, or lacking in, social understanding.

Interpersonal Reactivity Index (IRI) (Davis 1980): The IRI is purportedly an effective (Muncer and Ling 2006) and commonly used self-report empathy instrument (Alterman et al. 2003). It has been employed in recent human–animal interaction studies (e.g., Taylor and Signal 2005; Henry 2006) and has excellent reliability and validity. Participants indicate their responses on a 5-point Likert scale, ranging from "Does Not Describe Me Well" to "Describes Me Very Well." The 28 items comprise four subscales, each of which taps a different dimension of empathy. The four subscales are as follows: (1) Fantasy Scale (FS), which explores the respondent's inclination to identify with fictitious characters, such as those from books or movies; (2) Perspective Taking (PT), which measures the individual's ability to adopt another person's view; (3) Empathic Concern (EC), which assesses the respondent's ability to feel compassion for others who engaged in negative experiences; and (4) Personal Distress (PD), which indicates the extent to which individuals witness others' negative experiences, resulting in their own anxiety and discomfort. While this empathy instrument has been used in previous research in order to measure human-directed empathy (Miller and Eisenberg 1988; Alterman et al. 2003; Perez-Albeniz and de Paul 2003), Taylor and Signal (2005) have reported significant correlations between this instrument and attitudes toward animals (Empathic Concern scale, $r = 0.33$).

Participants and Procedures

The participants in this study were undergraduate students from a university in southwestern Ontario (Canada). While there were initially 424 (M = 136, F = 288) participants, incomplete data reduced the usable sample size to 387 (M = 119, F = 268). All were enrolled in a large, on-line, introductory writing class open to all undergraduate students. Participation was invited via the course's website, on which a notice was posted asking students to complete a voluntary survey for which they would receive a 1% bonus mark. Interested students were directed to the survey on the interactive website, Survey Monkey. The major portion of the sample (73%) was aged from 18 to 21 years; 16% were aged from 22 to 24 years, and, overall, the ages ranged from 17 to 52 years (Mean = 21.6 years; Median = 20, Mode = 18). The broad range of declared majors exceeded ten, including Sociology (19%), Psychology (16%), Business (15%), and Science (8%). The dog and cat ownership status of participants, currently and in their childhood, is given in Table 1.

Table 1. Previous and current dog and cat ownership of the participants.

	Pet Ownership in Childhood	Current Pet Ownership
Neither Cat nor Dog Owned	136 (31.9%)	209 (48.9%)
Both Cat and Dog Owned	130 (30.4%)	41 (9.6%)
Dog Only	108 (25.3%)	116 (27.2%)
Cat Only	53 (12.4%)	61 (14.3%)

Data Analysis

The data were analyzed using multivariate analysis of covariance (MANCOVA), and chi-square tests. The research question of interest was the empathic profile of participants as a function of pet ownership (1) related to dogs and/or cats, and (2) when the participants were children and adults. Four categories of pet ownership related to dogs and cats were determined: (1) Neither, (2) Dog-Only, (3) Cat-Only, and (4) Both (cat and a dog). Two separate analyses were run, one for the childhood ownership data and one for the adult ownership data. Sex was entered as a covariate to control for the propensity for females to score higher on empathy measures than males.

Results

Childhood Pattern

With respect to pet ownership as a child, a MANCOVA was run with Pet Category (Neither, Dog-Only, Cat-Only, Both) as the independent variable and performance on the eight profile measures as the dependent variables. Sex was entered as a covariate. There was a main effect for Pet Category in the omnibus test ($F_{(24, 1146)}$ = 1.87, $p < 0.01$). Subsequent univariate tests revealed differences on the IRI-Personal Distress scale ($p < 0.05$), on the EQ-Social Skills factor ($p < 0.01$), and on the AAS ($p < 0.01$). On the IRI-Personal Distress scale, post-hoc analyses revealed that the Neither group showed higher personal distress than the Dog-Only group ($p < 0.05$) and the Both group ($p < 0.01$). For the EQ-Social Skills factor, the post-hoc analyses showed lower ratings for those in the Neither group compared with the Dog-Only ($p < 0.01$) and the Both groups ($p < 0.01$). Regarding the AAS, the Neither group scored lower than all three pet groups: Dog-Only ($p < 0.01$), Cat-Only ($p < 0.05$), and Both ($p < 0.01$) (means and standard deviations are reported in Table 2). In summary, of particular interest here is that those who were dog owners as children (whether having a dog only or having both a dog and a cat) had lower personal distress and higher social skills.

Adult Pattern

With respect to pet ownership as an adult (current), the MANCOVA revealed a similar main effect for Pet Category in the omnibus test ($F_{(24, 1146)}$ = 3.03, $p < 0.001$). Subsequent univariate tests and post-hoc analyses indicated differences on the AAS, the EQ-Social Skills factor, and the IRI-Personal Distress scale, which were somewhat different from the childhood pattern (means and standard deviations are reported in Table 3). All three pet-owning groups (Cat-Only,

Table 2. Mean (and standard deviation) scores on the IRI scales, the EQ factors, and the Animal Attitude Scale, according to dog and cat ownership during the childhood of the participants.

	Dog or Cat Owned?			
	Neither (n = 124)	Dog-Only (n = 99)	Cat-Only (n = 51)	Both (n = 118)
IRI-Perspective Taking	3.46 (0.68)	3.55 (0.62)	3.42 (0.69)	3.45 (0.63)
IRI-Fantasy Scale	3.26 (0.71)	3.33 (0.79)	3.36 (0.82)	3.36 (0.76)
IRI-Emotional Concern	3.60 (0.44)	3.58 (0.56)	3.55 (0.50)	3.60 (0.44)
IRI-Personal Distress	2.88 (0.65)	2.70 (0.63)	2.78 (0.67)	2.69 (0.62)
EQ-Cognitive Empathy	3.01 (0.64)	3.07 (0.59)	3.05 (0.48)	3.13 (0.55)
EQ-Emotional Reactivity	2.94 (0.64)	3.03 (0.59)	2.93 (0.51)	3.11 (0.56)
EQ-Social Skills	2.79 (0.64)	3.05 (0.54)	2.84 (0.44)	2.99 (0.54)
Animal Attitude Scale	3.24 (0.51)	3.48 (0.68)	3.41 (0.60)	3.48 (0.61)

Dog-Only, Both) scored higher on the AAS than the Neither group ($p < 0.01$); the Both group scored higher than the Cat-Only and Dog-Only groups ($p < 0.01$) also. As can be seen in Figure 1, this was different from the childhood pattern. Personal Distress in the Neither group was higher but only when compared with the Dog-Only group ($p < 0.05$). Like the childhood pet ownership data, there was a difference on the EQ-Social Skills factor for adult owners: the Dog-Only group scored higher than the Neither group ($p < 0.01$). In addition, the Dog-Only group scored higher than the Cat-Only group ($p < 0.025$).

Table 3. Mean (and standard deviation) scores on the IRI scales, the EQ factors, and the Animal Attitude Scale, according to current dog and cat ownership of the participants.

	Dog or Cat Owned?			
	Neither (n = 192)	Dog-Only (n = 105)	Cat-Only (n = 58)	Both (n = 37)
IRI-Perspective Taking	3.48 (0.62)	3.52 (0.70)	3.37 (0.70)	3.46 (0.55)
IRI-Fantasy Scale	3.24 (0.71)	3.33 (0.82)	3.53 (0.74)	3.41 (0.81)
IRI-Emotional Concern	3.56 (0.47)	3.62 (0.48)	3.56 (0.50)	3.65 (0.49)
IRI-Personal Distress	2.79 (0.66)	2.67 (0.60)	2.77 (0.66)	2.86 (0.65)
EQ-Cognitive Empathy	3.04 (0.57)	3.09 (0.58)	3.02 (0.71)	3.23 (0.35)
EQ-Emotional Reactivity	2.97 (0.57)	3.06 (0.60)	2.91 (0.72)	3.25 (0.36)
EQ-Social Skills	2.86 (0.54)	3.07 (0.60)	2.86 (0.70)	2.95 (0.30)
Animal Attitude Scale	3.21 (0.47)	3.54 (0.64)	3.48 (0.74)	3.82 (0.55)

Child–Adult Combinations

A crosstabs analysis was run to examine differences between child patterns and adult patterns more closely. As may be recalled from Table 1, there were differences in the child and adult ownership patterns: these were statistically significant ($\chi^2_{(9)} = 2.83$, $p < 0.001$). To illustrate, of the 136 participants who had neither a cat nor a dog as a child, 121 showed the same pattern as an adult (89%). Of the 130 participants who had both a dog and a cat as a child, only 37 (28.5%) had both a dog and a cat as an adult. With respect to the Dog-Only group, 34.3%

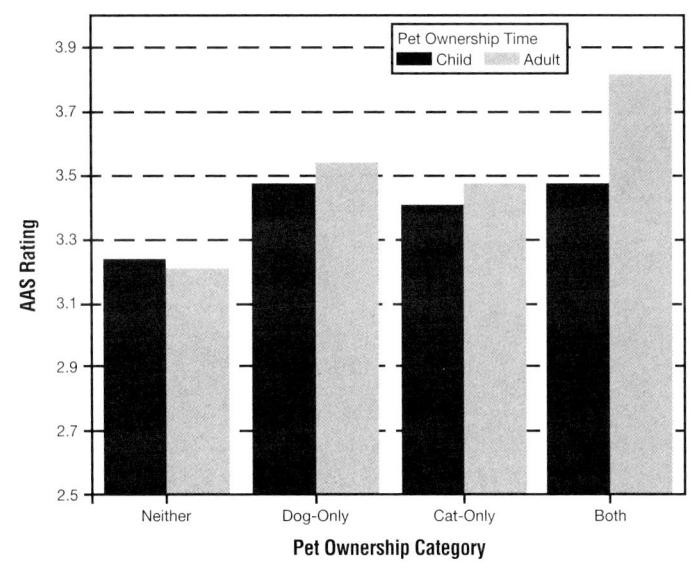

Figure 1. Mean Animal Attitude Scale scores for different types of pet ownership during childhood and adulthood.

of those who only had a dog as a child also only had a dog as an adult. For the Cat-Only group, 41.5% of those who only had a cat as a child also only had a cat as an adult. Clearly there is some diversity in child and adult pet ownership patterns. Unfortunately, the frequency counts for many of the cells were too small to permit more sophisticated statistical analyses of the total group.

To consider the differences graphically, the figures to follow are structured to illustrate where differences occur. In Figure 1, we see the low scores on the Animal Attitude Scale for those participants, both in childhood and adulthood, who owned neither a cat nor a dog. But what is more striking in the figure is the dramatically higher score for adults with respect to owning both a cat and a dog. Of the 130 participants who had both a dog and a cat as a child, only 37 had both as adults (28.5%); however, those 37 adults who had both animals represent over 90% of all those who had both as adults (37/41). The findings suggest an emergent positive attitude by adulthood.

The Both group, in adulthood, shows high Personal Distress, much like the Neither group, but this difference is not statistically significant, perhaps due to the smaller cell size for the comparisons (Both = 37; Neither = 192). The significant difference is between the Neither group and the Dog-Only group. Thus, what Figure 2 does show is the lower Personal Distress for the Dog-Only group (childhood and adulthood).

Figure 3 is interesting in that it shows the Dog-Only group scoring higher on the EQ-Social Skills factor. The effect holds for both the child pattern ($n = 108$) and the adult pattern ($n = 116$), regardless of an overlap of only 60.2%. That is, only 65 of the participants who only owned a dog as a child, also only owned a dog as an adult. But for the group who fell into the Dog-Only category—whether during childhood or adulthood—the pattern of enhanced social skills was pronounced.

To summarize, the Dog-Only group scored higher on the EQ-Social Skills factor, a response, which holds for both child and adult patterns of pet ownership (see Figure 3).

Empathic Differences in Adults as a Function of Childhood…

Secondly, those who owned both dogs and cats, or only dogs, as children, scored lower on the IRI-Personal Distress scale, a pattern which is evident for adults as well, but only for those who only owned dogs (see Figure 2). And thirdly, those who owned dogs and cats as children and adults had higher scores on the AAS than those who reported not having dogs or cats as pets (see Figure 1).

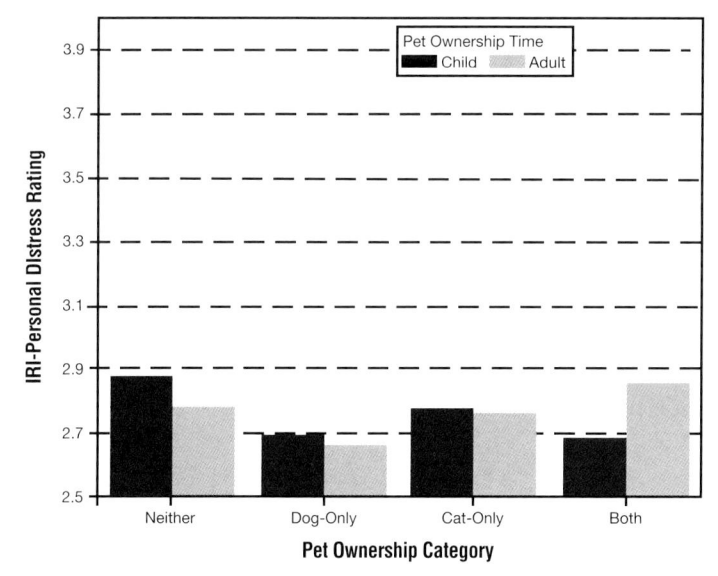

Figure 2. Mean IRI-Personal Distress scores for different types of pet ownership during childhood and adulthood.

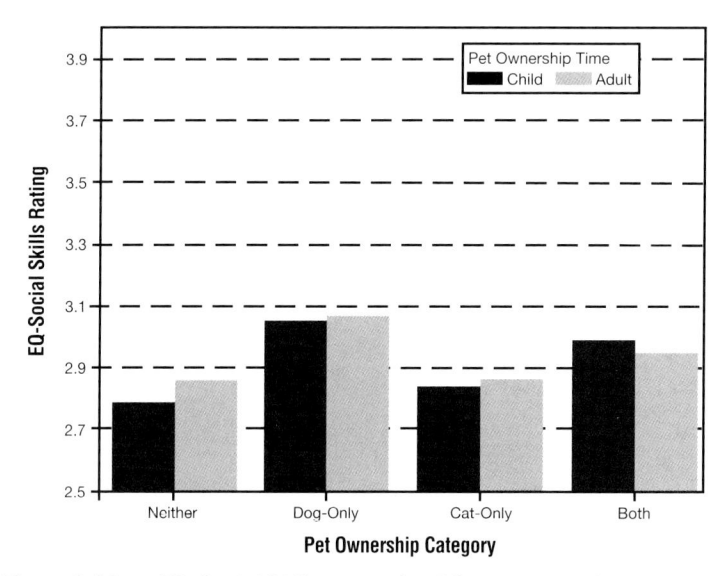

Figure 3. Mean EQ-Social Skills scores for different types of pet ownership during childhood and adulthood.

Discussion
Empathy Quotient: Social Skills Factor
For the EQ-Social Skills factor, adults who only had dogs in childhood scored higher than those who only had a cat or who did not have any pets. This resonates with previous research: Daly and Morton (2003) reported that children who owned dogs were higher in general empathy than those who owned cats or those who did not have any pets. In the present study, individuals who reported only having dogs as adults, as well as only having had dogs as children, scored higher on the EQ-Social Skills factor than both the Cat-Only and the Neither groups. By definition, those who score highly on this scale have a heightened ability to spontaneously employ skills as they relate intuitively to social understanding (Lawrence et al. 2004). These findings of a dog-related social-skills effect for both adult and child groups were not surprising, given the wealth of literature indicating that dogs positively contribute to the social environment of both adults and children and in various circumstances. For instance, Mader, Hart and Bergin (1989) found that children in wheelchairs received more social acknowledgement when they appeared in public with dogs. Similarly, McNicholas and Collis (2000) reported that dogs served as social catalysts for interactions with strangers for adults who appeared with dogs in public. A study involving a female participant accompanied at various times by one of three types of dog (a Labrador puppy, an adult Labrador, and an adult Rottweiler), two neutral objects (a teddy bear and a plant), or alone, found that while interactions with strangers were higher for both the puppy and the adult Labrador, dogs in general elicited a highly significant number of social interactions (Wells 2004). In a recent study (Esteves and Stokes 2008), children with developmental disabilities increased positive interactions with their teachers when a dog was present, and the researchers observed an increase in the general social responsiveness of all students in the classroom.

With respect to the social skills aspect of empathy in the present study, it appears that ongoing relationships with dogs is instrumental, and the impetus begins in childhood, at least in part. Previous research lends support to the notion that pet owning in childhood may affect empathy in adolescence and adulthood. Vizek-Vidović et al. (2001) conducted research, partly retrospective in nature, examining the quality of relationships that adults had with their childhood pets, and had similar findings to the present investigation: adults who had had pets (regardless of type) in childhood were reportedly more empathic in late adolescence and adulthood. They also noted that ownership in childhood may be related to choosing careers in a "helping profession," suggesting that perhaps childhood pet ownership not only facilitates empathy development, but "socially oriented values" including social interaction.

Pertinent to our current findings, however, there was a difference between those who specifically had dogs as children and those who had no pet (not even cats), similar to the findings of our previous work (Daly and Morton 2003). The logical flow of our current rationale points to the absence of a pet being equal to the absence of a social facilitator. But this does not explain the intriguing finding that cat ownership was comparable to not having had a pet at all. A reasonable and intuitively sensible rationale is related to the nature of one's relationship with dogs, a relationship that logically facilitates social empathy due to the nature of the human–dog relationship. Bonas, McNicholas and Collis (2000) showed that the human–pet relationship is generally characterized as a type of social relationship. Seemingly rooted in humans' anthropomorphic tendencies to interact and communicate with their pets as they do with other humans, individuals in this study consistently regarded their pets as "significant others" and family members, with relationships with dogs generally being significantly stronger than those with cats. It might also be reasonably posited that individuals with high social empathy naturally gravitate towards the dog as pet.

Interpersonal Reactivity Index: Personal Distress Factor
Our second notable finding, and equally interesting, was that those who as children had neither dogs nor cats scored higher on the IRI-Personal Distress scale, when compared with those who only had dogs or had both dogs and cats. It seems that having pets in childhood (dogs, or both dogs and cats) relates to improved mental health with respect to personal distress. This effect seems to carry over to adulthood, in part, in that adults who were in the Dog-Only group had lower IRI-Personal Distress scale scores than those in the Neither group. Can dogs have a beneficial effect on mental health to the point of reducing personal distress? It seems this is a possibility worth further exploration.

How dogs might reduce personal distress is not clear. But consider that Fidler, Light and Costall (1996) reported that individuals who had a history of experience with dogs and/or cats were able to describe the animals in terms of feelings, etc. Bonas, McNicholas and Collis (2000) found that, overall, human–pet relationships provided support comparable to human–human relationships. This would seem to involve both cognitive (description) and affective (tapping into feelings) aspects, and might also be regarded as consistent with pet owners' inclination to anthropomorphize their pets (a phenomenon described by Serpell (2003) as "the attribution of human states … to nonhuman animals"). In recent years, anthropomorphizing pets has increased in popularity, raising questions about the nature of this behavior (Antonacopoulos and Pychyl 2008). In a recent study reporting that individuals found more security in their relationships with pets than with romantic partners, Beck and Madresh (2008) suggested that pets provide a more consistent and uncomplicated role in the lives of their owners than do romantic partners. This could lead to reduced personal distress.

There is literature suggesting that many pet owners enjoy relationships with their dogs and cats that are akin to those with humans, which may well explain the low scores on the IRI-Personal Distress scale for the individuals who currently owned dogs. Given the relatively young age of this sample group, they are less likely to have their own children, and as such, may regard pets as child substitutes. At the very least, the finding warrants further investigation into the nature of this phenomenon and the unclear emotional distinction, as suggested elsewhere (Bonas, McNicholas and Collis 2000; Beck and Madresh 2008), between human–animal and human–human relationships.

Animal Attitude Scale
Here our final findings were: (1) logical, in that animal attitude scores were more positive for those who had pets as children and/or as adults, and (2) interesting, in that adults who owned both dogs and cats ($n = 41$) demonstrated more positive attitudes toward animals than did those who owned both pets as children ($n = 130$). This is not too surprising considering that of the 41 adults who owned both cats and dogs, 37 of them also owned both as children (i.e., 90%). It is a select group of adults, apparently attached to both dogs and cats as children and adults, which showed the enhanced attitude.

Limitations
These ownership patterns need to be interpreted with caution and certain caveats. For instance, the majority (89%) of the sample group was aged from 18 to 24 years, making it unreasonable to come to specific conclusions regarding adult pet ownership in general. In the

same vein, it should also be noted that because the participants in this study were university students, they were likely to be living in a transient state, for example, with roommates, away from home, in an apartment, or temporarily living in the city. For this reason, it may not be surprising that many with specific childhood pets showed a different adult pattern.

Conclusion

There is a case to be made that pet ownership experiences and patterns as children impact on adult ownership patterns for a large segment of adults. More importantly, there is a case to be made that pet ownership patterns impact on personality variables (such as personal distress and attitudes) and personal behaviors (such as social skills). Such effects warrant careful, further investigation. Future research endeavors should also include samples of older adults, such as those of middle and senior ages, in order to glean a more definitive examination of patterns of pet ownership throughout adult life across the spectrum.

References

Alterman, A. I., McDermott, P. A., Cacciola, J. S. and Rutherford, M. J. 2003. Latent structure of the Davis Interpersonal Reactivity Index in methadone maintenance patients. *Journal of Psychopathology and Behavioral Assessment* 25: 257–265.

Antonacopoulos, N. M. D. and Pychyl, T. A. 2008. An examination of the relations between social support, anthropomorphism and stress among dog owners. *Anthrozoös* 21: 139–152.

Ascione, F. R. 1992. Enhancing children's attitudes about the humane treatment of animals: Generalization to human-directed empathy. *Anthrozoös* 5: 176–191.

Ascione, F. R. and Weber, C. V. 1996. Children's attitudes about the humane treatment of animals and empathy: One-year follow up of a school-based intervention. *Anthrozoös* 9: 188–195.

Baron-Cohen, S. and Wheelwright, S. 2004. The empathy quotient: An investigation of adults with Asperger syndrome or high functioning autism, and normal sex differences. *Journal of Autism and Developmental Disorders* 34: 163–175.

Beck, L. and Madresh, E. A. 2008. Romantic partners and four-legged friends: An extension of attachment theory to relationships with pets. *Anthrozoös* 21: 43–56.

Bonas, S., McNicholas, J. and Collis, G. 2000. Pets in the network of family relationships: An empirical study. In *Companion Animals and Us: Exploring the Relationships between People and Pets*, 209–236, ed. A. L. Podberscek, E. S. Paul and J. Serpell. Cambridge: Cambridge University Press.

Daly, B. and Morton, L. L. 2003. Children with pets do not show higher empathy: A challenge to current views. *Anthrozoös* 16: 298–314.

Daly, B. and Morton, L. L. 2006. An investigation of human–animal interactions and empathy as related to pet preference, ownership, attachment, and attitudes in children. *Anthrozoös* 19: 113–127.

Daly, B. and Morton, L. L. 2008. Empathic correlates of witnessing the inhumane killing of an animal: An investigation of single and multiple exposures. *Society & Animals* 16: 243–255.

Davis, M. H. 1980. A multidimensional approach to individual differences in empathy. *JSAS Catalog of Selected Documents in Psychology* 10: 85.

Esteves, S. W. and Stokes, T. 2008. Social effects of a dog's presence on children with disabilities. *Anthrozoös* 21: 5–15.

Fidler, M., Light, P. and Costall, A. 1996. Describing dog behavior psychologically: pet owners versus non-owners. *Anthrozoös* 9: 196–200.

Fifield, S. J. and Forsythe, D. K. 1999. A pet for the children: Factors related to family pet ownership. *Anthrozoös* 12: 24–32.

Friedmann, E., Locker, B. Z. and Lockwood, R. 1993. Perception of animals and cardiovascular responses during verbalization with an animal present. *Anthrozoös* 6: 115–134.

Friedmann, E. and Thomas, S. A. 1995. Pet ownership, social support, and one-year survival among post-myocardial infarction patients in the cardiac arrythmia suppression trial (CAST). *American Journal of Cardiology* 76: 1213–1217.

Friedmann, E., Thomas, S. A., Cook, L. K., Tsai, C.-C. and Picot, S. J. 2007. A friendly dog as potential moderator of cardiovascular response to speech in older hypertensives. *Anthrozoös* 20: 51–63.

Grier, K. C. 1999. Childhood socialization and companion animals: United States, 1820–1870. *Society & Animals* 7: 95–120.

Henry, B. C. 2004. The relationship between animal cruelty, delinquency, and attitudes toward the treatment of animals. *Society & Animals* 12: 185–207.

Henry, B. C. 2006. Empathy, home environment, and attitudes toward animals in relation to animal abuse. *Anthrozoös* 19: 17–34.

Hergovich, A., Monshi, B., Semmler, G. and Zieglmayer, V. 2002. The effects of the presence of a dog in the classroom. *Anthrozoös* 15: 37–50.

Herzog, H. A., Betchart, N. S. and Pittman, R. B. 1991. Gender, sex role orientation, and attitudes toward animals. *Anthrozoös* 4: 184–191.

Lawrence, E. J., Shaw, P., Baker, D., Baron-Cohen, S. and David, A. S. 2004. Measuring empathy: Reliability and validity of the Empathy Quotient. *Psychological Medicine* 34: 911–924.

Levinson, B. M. 1978. Pets and personality development. *Psychological Reports* 42: 1031–1038.

Mader, B., Hart, L. and Bergin, B. 1989. Social acknowledgements for children with disabilities: Effects of service dogs. *Child Development* 60: 1529–1534.

McNicholas, J. and Collis, G. M. 2000. Dogs as catalysts for social interaction: Robustness of the effect. *British Journal of Psychology* 91: 61–70.

Melson, G. F. 2003. Child development and the human–companion animal bond. *American Behavioral Scientist* 47: 31–39.

Melson, G. F., Peet, S. and Sparks, C. 1992. Children's attachment to their pets: Links to socio-emotional development. *Children's Environments Quarterly* 8: 55–65.

Miller, P. A. and Eisenberg, N. 1988. The relation of empathy to aggressive and externalizing/antisocial behavior. *Psychological Bulletin* 103: 324–344.

Muncer, S. J. and Ling, J. 2006. (Human-ANIMAL) Psychometric analysis of the Empathy Quotient (EQ) scale. *Personality and Individual Differences* 40: 1111–1119.

Paul, E. S. 2000a. Empathy with animals and with humans: Are they linked? *Anthrozoös* 13: 194–202.

Paul, E. S. 2000b. Love and pets and love of people. In *Companion Animals & Us: Exploring the Relationships Between People and Pets*, 168–186, ed. A. L. Podberscek, E. S. Paul and J. A. Serpell. Cambridge: Cambridge University Press.

Paul, E. S. and Serpell, J. A. 1993. Childhood pet keeping and humane attitudes in young adulthood. *Animal Welfare* 2: 321–337.

Perez-Albeniz, A. and de Paul, J. 2003. Dispositional empathy in high- and low-risk parents for child physical abuse. *Child Abuse and Neglect* 27: 769–780.

Poresky, R. H. and Hendrix, C. 1990. Differential effects of pet presence and pet-bonding on young children. *Psychological Reports* 67: 51–54.

Serpell, J. 2003. Anthropomorphism and anthropomorphic selection—beyond the "cute response." *Society & Animals* 11: 437–454.

Signal, T. D. and Taylor, N. 2007. Attitude to animals and empathy: Comparing animal protection and general community samples. *Anthrozoös* 20: 125–130.

Taylor, N. and Signal, T. D. 2005. Empathy and attitudes to animals. *Anthrozoös* 18: 18–27.

Vizek-Vidović, V., Arambasic, G. K., Kerestes, G., Kuterovac-Jagodic, G. and Vlahovic-Štetić, V. 2001. Pet ownership in childhood and socio-emotional characteristics, work values, and professional choices in early adulthood. *Anthrozoös* 14: 224–231.

Vizek-Vidović, V., Štetić, V. V. and Bratko, D. 1999. Pet ownership, type of pet, and socioemotional development of school children. *Anthrozoös* 12: 211–217.

Wells, D. L. 2004. The facilitation of social interactions by domestic dogs. *Anthrozoös* 17: 340–352.

Social and Individual Components of Animal Contact in Preschool Children

Manuela Wedl and Kurt Kotrschal

Konrad Lorenz Research Station and Department of Behavioural Biology, University of Vienna, Austria

Address for correspondence:
Manuela Wedl, Department of Behavioural Biology, University of Vienna, Althanstraße 14, A-1090 Vienna, Austria. E-mail: manuela.wedl@univie.ac.at

ABSTRACT Humans are generally biophilic. Still, for unknown reasons, interest in animals varies substantially among individuals. Our goal was to investigate how differential interest of children towards animals might be related to social competence and personality. We proposed two alternatives: 1) Children may compensate for potential deficits in social competence by resorting to animals, and 2) Socially well-connected children may show a particular interest in animals. We focused on relationships between age, gender, family background, play behavior, personality components, and contact with rabbits in 50 children (22 boys/28 girls; 3 to 7 years of age) at a preschool in Krems/Austria. Data were analyzed using GLM. We found that each one of these variables had significant impact on intensity of engagement with the rabbits. In general, girls, children with siblings, and children without pets were more oriented towards the rabbits than were boys, children without siblings, or pet-owning children. The older the children, the less frequently they occupied themselves with the rabbits but the longer they remained when they did engage them. Furthermore, we found that the more "Confident/Respected" (PCA factor 1) and less "Patient/Calm," "Cheerful/Sociable," and "Solitary" (PCA factors 2–4) the children, the more time they spent in direct occupation with rabbits. Most effects of the investigated variables varied between boys and girls. By and large, our findings support the hypothesis that the "socially competent" children were particularly interested in the animals. Also, children's social styles, as evinced in interactions with peers, were generally reflected in how they interacted with the rabbits.

Keywords: companion animals, human–animal interactions, personality, pets, rabbits

❖ Contact with animals may positively affect human health and well-being (e.g., Katcher and Beck 1985; Collis and McNicholas 1998; Friedmann 2000), notably through the ability of animals to provide social support, to physically activate their human companions (Bachmann 1975; Brickel 1982), and to increase their social contact with other humans (Mugford and M'Comisky 1975; Messent 1983; McNicholas and Collis 2000). Companion animals also may increase the social status of children within their peer groups (Guttmann, Predovic and Zemanek 1985). The presence of a dog in a classroom increased children's attentiveness and social cohesion (Kotrschal and Ortbauer 2003), and children in a classroom with a dog present were shown to have developed greater self-confidence and social competence than children in a control group without a dog (Hergovich et al. 2002). Such results support pedagogical common sense that animal contact benefits child development. In fact, an increasing number of institutions have been integrating animals into their programs.

This was the case in a preschool in Krems (Austria), where 50 children between 3 and 7 years of age had free access to rabbits during a specified period every day. This situation allowed us to tackle the question: Why, despite the general human tendency toward biophilia (Wilson 1984), does interest in animals vary quantitatively and qualitatively among individuals? Whether, and to what extent, humans enjoy the company of animals is a complex matter (Serpell 1986). Although animal lovers are sometimes seen as indulging their interest at the expense of their regard for fellow humans, this does not actually seem to be the case (Paul 1995; 2000). Our present study of preschool children may have the potential to shed more light onto the question of how interest in animals is related to social and individual parameters. Our goal was to investigate whether and how differential demonstration of interest in animals could be related to age, gender, family background, play behavior, and personality components—variables we considered to be the most important factors in children of this age group.

We hypothesized that the presence of animals could compensate for individual deficits in social connectivity (social compensation hypothesis), or, conversely, that socially competent children would be particularly interested in animals (Paul 2000; social competence hypothesis) and would therefore seek more contact with the rabbits.

Methods
Participants
We studied 50 children (28 girls and 22 boys, 3 to 7 years of age; Table 1) at a preschool in Krems/Austria, operated by a staff of seven female teachers and assistants. Children were not divided into groups assigned to specific rooms or spaces; all were able to utilize the entire premises, which offered a range of possibilities for group play and solitary play.

Table 1. Gender and age of children as related to year of attendance at the preschool.

Year of Attendance at the Preschool	Number of Children	Gender (*n* female/*n* male)	Mean Age (years)	Age Range (years)
1st	17	9/8	4.17	3.3–4.6
2nd	17	10/7	5.15	4.8–5.6
3rd	16	9/7	6.19	5.7–7.0

Setting
During the observation period, one large, adult, male rabbit occupied a cage in the garden and two smaller adult, female rabbits along with three young were housed in two hutches in the preschool building's entrance area. The male rabbit was regularly released into a larger enclosure in the garden and the other rabbits were allowed to move freely in the entrance area. All children had free, ad libitum access to the rabbits during free playtime from 0700 until 1200

hours, except during daily programs (generally 0830–0845 hours, 1000–1100 hours). They had been instructed how to treat the rabbits appropriately and mutual social control enforced these rules. No serious mishandling or abuse of rabbits was ever observed.

Procedure

As a matter of principle, our research was as "minimally invasive" as possible; that is, none of the usual routines in the preschool was changed or significantly adapted for data collection. Parents were informed about the study in a parent–teacher conference and were given the opportunity to discuss our research plan and to contribute to it. Consent from all parents was obtained, as were permissions from the relevant governmental offices. A pilot study, conducted from February through April 2004, served to test and improve our procedures and to habituate the children to the observer's (first author) presence, to videotaping, and direct observation. No data from this pilot were used in the current analysis; the primary data were collected in May and June 2004. When the preschool's schedule differed from the usual daily routine, no data were taken (e.g., because of special activities, birthdays and other celebrations, visits from external preschool teachers or children, visits from trainees, trips, "open house," etc.).

Child–Animal Interactions

In the entrance area, one video camera (Sony DCR-TRV 19 E) equipped with a wide-angle lens was installed on the wall. In the garden, a second video camera (JVC GR-DVL 145) was placed on a tripod at a distance of approximately 10 meters from the rabbit enclosure. Video recordings were conducted in the mornings between 0700 and 1200 hours. Child–rabbit interactions during 9 days distributed between May 24 and June 23, 2004 were coded from these videotapes. Inter-coder reliability was tested in several video sections throughout the coding process. The percentage of agreement (duration; The Observer 5.0 software, Noldus Information Technology, The Netherlands) was 80%; as measured by calculating Cohen´s Kappa, inter-coder reliability was 0.72.

Tapes were behavior-coded by the first author using The Observer 5.0. Because many children passed through the foyer observation area for reasons unrelated to interest in the rabbits, we included only those children who were occupied with the animals at least once while in the observation field. Coding started when the child entered the field and ended when he/she left. During the 9 days coded, each of the 50 children attended preschool for 4 to 9 mornings. The male rabbit was released into the garden enclosure on 7 of the 9 days, and one or more of the rabbits housed in the entrance area were taken out of their hutches on 5 of the 9 days.

All relevant behaviors shown by the children while focused on, and interacting with, the rabbits were coded via continuous recording. For statistical analysis, the behaviors (Table 2) were later grouped into the following categories: "direct occupation with rabbits" (one or more rabbits directly at the focus of attention/action) and "indirect occupation with rabbits" (focus of occupation is only indirectly related to rabbits). The category "proximity to rabbits" (child's presence apparently unrelated to rabbits but in connection with prior or subsequent occupation with them) was also used for analysis. We used the terms "occupy" and "occupation" to encompass the various ways children were observed to exhibit and act on their interest in rabbits. Because not all children were present in the preschool during all observation days, data were standardized with respect to observation effort (per observation day). The frequency of being occupied with rabbits and the duration of the three categories of occupation (direct and indirect occupation with rabbits and proximity to rabbits) were used for statistical analysis.

Table 2. Specific behaviors constituting direct and indirect occupation with rabbits and definitions.

Behavior	Definition
Watch Rabbits [a]	Child observes one or more rabbits without direct or indirect engagement with them.
Watch Persons in Contact with Rabbits [a]	Child observes interaction between one or more persons and one or more rabbits without being directly engaged.
Stroke Rabbit [a]	Child passes one or both hands over a rabbit's fur.
Hold Rabbit [a]	Child holds a rabbit in his/her arms or on lap.
Feed Rabbits [a]	Child places food into rabbit enclosure or hutch or feeds rabbit(s) by hand.
Reach into Hutch [a]	Child reaches into rabbit hutch without touching a rabbit.
Open or Close Hutch Door [a]	Child handles hutch in the act of opening or closing the door.
Follow Rabbits [a]	Child moves after one or more rabbits.
Look for Rabbits [a]	Child looks around or behind furniture in entrance area or enclosure.
Talk to Rabbit [a]	Child speaks to rabbit while simultaneously orienting his/her face toward rabbit.
Take Rabbit Out of, or Puts into, Cage [a]	Child picks up rabbit and removes it from or puts it into cage.
Touch Rabbit [a]	Child contacts rabbit but does not stroke or hold rabbit.
Chase Rabbit [a]	Child rapidly pursues rabbit.
Drop Rabbit [a]	Child lets rabbit fall or pushes it.
Talk about Rabbits [b]	Child participates in a conversation about rabbits.
Look through Rabbit Books [b]	Child looks through one or more rabbit books.
Pick Fresh Fodder [b]	Child forages, picking greens for the rabbit in the garden.
Handle Rabbit Cage/ Enclosure Material [b]	Child touches or works with a cage, the enclosure fence or ladder, or the water trough.
Imitate Rabbit [b]	Child copies behavior of rabbit.
Clean Cage Area [b]	Child assists preschool staff in cleaning; e.g., helps with dustpan or broom.
Watch Cleaning of Cage Area [b]	Child watches someone cleaning the area around a cage.

[a] Denotes direct occupation with rabbits.   [b] Denotes indirect occupation with rabbits.

Social Interactions with Other Children and Personality

Data on child play behavior were scored via direct observation during free playtime (~ 0815 to 0845 hours) throughout the entire premises, except entrance area and garden, for a total of 17 days between May 3 and June 9, 2004. At 10-minute intervals, the observer (first author) moved in a designated order through all the rooms to which children had access. Observations were conducted by visually scanning from right to left. Check sheets were used to record whether a child was playing in a group, playing in parallel to other children, playing alone, or watching the activities of other children (after Grammer 1995). Each child was observed 12 to 38 times during this period (a total of 1,413 play events). The number of children observed per scan varied because not all children were present on all days and because not all children present were visible in each scan. Therefore, percentages (from the total number of observations) of group play, parallel play, solitary play, and watching other children were calculated for each child. Inter-observer reliability was tested using sections from video recordings made with a hand-held video camera during the pilot study. Only events in which the children were fairly visible and/or audible were selected. Observations were conducted from the selected video sections by use of check sheets. Inter-observer agreement on observed child play behavior was 97.9%.

Two preschool teachers independently rated the children on 12 personality items (5-point Likert scale, modified after Seabrook 1984 and Waiblinger 1996). Also, on a 5-point scale (from high to low), they rated the children's social status according to how respected each child was by their peers. We used a list of criteria, modified from Hold (1976), as a guide for rating the children on this scale. In most cases, we calculated the mean value of the independent personality ratings of the two preschool teachers. For items where the difference between the ratings of the two observers was greater than 2 on the 5-point scale (in only 12 out of 650 cases), we discarded the rating and substituted it with the rating of the first author and used the mean of this and the retained rating. In other questionnaires, we asked about pet ownership and sibling status of the children.

A Principal Component Analysis (PCA; Bartlett-Test: KMO = 0.726; Sphericity: χ^2 = 732.695, df = 136, p < 0.001) with varimax rotation of data on child play behavior and on the rated personality items revealed four major components: "Confident/Respected," "Patient/Calm," "Cheerful/Sociable," and "Solitary" (Table 3, p. 388). These explained 77% of the variance.

Analysis
Data analysis was carried out with the aid of The Observer 5.0. and SPSS 15.0 (Chicago, IL, USA). We conducted four generalized linear models (GLMs) with a log-link function and based on Poisson distribution.

The first GLM was constructed with "frequency of occupation with rabbits" as the response variable (dependent variable); "gender," "siblings yes/no," "pets yes/no" as factors; and "age," "PCA factor 1," PCA factor 2," PCA factor 3" and "PCA factor 4" as covariates. We selected these explanatory variables as main effects and included interactions between gender and each of the other explanatory variables. These interactions were chosen because other studies have reported gender differences in human–animal relationships (e.g., Ray 1982; Paul and Serpell 1992; Rost and Hartmann 1994; Herzog 2007), and we expected that the other explanatory variables would have different effects on the intensity of occupation with rabbits, depending on whether the child involved was a boy or a girl. We conducted pairwise comparisons with Bonferroni-correction to determine the differences between boys with siblings and without, between boys with pets at home and without, and the same for girls. The other three GLMs were constructed with "duration of direct occupation with rabbits," "duration of indirect occupation with rabbits," and "duration of proximity to rabbits" as response variables.

In all four models, we removed explanatory variables in order of decreasing significance (if p > 0.1). Only terms with p < 0.1 remained in the final model. Excluded terms were re-entered one by one into the final model to confirm that they did not explain a significant part of the variation (Poesel et al. 2006). Non-significant terms are not presented below. For the significant terms, we present Wald statistics.

Results
Gender Effects
Girls were occupied with the rabbits more often, and spent more time with them (direct and indirect occupation with rabbits and staying in proximity to rabbits), than the boys (Table 4 and Figure 1a–d).

Table 3. Results of Principal Component Analysis on data of child play behavior and personality items. The values represent the loading of each variable on the factors. Loadings of 0.500 or above are in bold.

Items Scored by Preschool Teachers on a 5-Point Scale and Data on Play Behavior	1: Confident/ Respected	2: Patient/ Calm	3: Cheerful/ Sociable	4: Solitary
Meek–Not Meek	**0.844**	–0.184	0.034	–0.023
Lacking Confidence–Confident	**0.843**	0.085	0.340	–0.074
Low Social Status Among Peers–High Social Status Among Peers	**0.810**	0.093	0.131	–0.195
Suspicious of Change–Liking Change	**0.800**	0.099	0.421	–0.029
One Who Keeps Quiet–One Who Speaks One's Mind	**0.794**	–0.288	0.323	0.021
Giving in Easily–Forceful	**0.792**	–0.346	0.111	0.095
Unsociable–Sociable	**0.673**	–0.092	**0.579**	–0.138
Giving up Easily–Persevering	**0.642**	**0.534**	0.096	0.171
Not Talkative–Talkative	**0.600**	–0.336	**0.514**	0.126
Impatient–Patient	–0.113	**0.887**	0.039	–0.017
A Worrier–Not a Worrier	–0.276	**0.858**	–0.025	–0.055
Difficult to Get on with–Easy-Going	0.423	**0.566**	**0.565**	0.114
Grumpy–Cheerful	0.315	0.121	**0.851**	–0.074
% of Group Play	0.000	–0.043	–0.034	**–0.989**
% of Watching Other Children	–0.255	–0.144	–0.176	**0.722**
% of Parallel Play	0.481	0.315	0.031	**0.534**
% of Solitary Play	0.020	–0.021	0.476	**0.533**

Children with Siblings

In general, children with siblings were occupied with the rabbits significantly more often, and spent more time with them (direct and indirect occupation with rabbits and staying in proximity to rabbits), than were only-children (Table 4 and Table 5). Three models revealed significant interactions with gender (Table 4). Both boys and girls with siblings were more often occupied with the rabbits, and were directly occupied with them for longer periods of time, than were only-children (Post-hoc test: all $p < 0.001$), but the difference in frequency was greater between boys with and without siblings, whereas the difference in duration of direct occupation was greater between girls with and without siblings (Table 5). Furthermore, girls with siblings spent significantly longer periods of time in proximity to the rabbits than girls without, whereas there was no significant difference between boys with siblings and without (Post-hoc test: girls: $p < 0.001$, boys: $p = 1.000$; Table 5).

Children with Pets of Their Own

Children without pets at home were in general significantly more often occupied with the rabbits, and spent significantly more time with them (direct and indirect occupation with the rabbits and staying in proximity to the rabbits), than were pet-owning children (Table 4 and Table 6). All four models revealed significant interactions with gender (Table 4). Girls without pets were occupied with the rabbits more often, and spent more time directly occupied with them, than were pet-owning girls, but no such differences were found in boys (Post-hoc test:

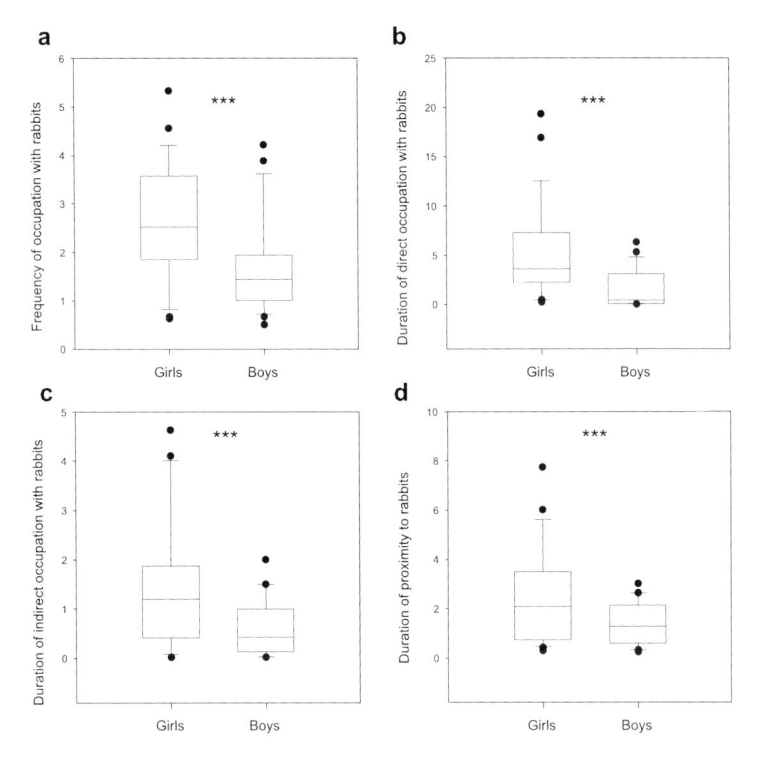

Figure 1. Gender differences in (a) mean frequency of occupation with rabbits (mean number of entrances into observation range followed by engagement with rabbits) per child per day, (b) mean duration of direct, (c) indirect occupation with rabbits, and (d) proximity to rabbits [min/child/day]; girls: $n = 28$, boys: $n = 22$. Box and whisker plots represent medians, upper and lower quartiles, and maximum and minimum scores. Dots indicate outliers. Asterisks mark significant differences, ***$p < 0.001$.

frequency: girls: $p < 0.001$, boys: $p = 1.000$; direct occupation: girls: $p < 0.001$, boys: $p = 0.824$; Table 6). In both boys and girls, children without pets at home spent more time indirectly occupied with the rabbits, and in proximity to them, than did pet-owning children (Post-hoc test: all $p < 0.001$), but the difference was greater in girls (Table 6).

Age Effects

In general, the older the children, the less often they were occupied with the rabbits (Table 4 and Figure 2a) and the less time they spent in proximity to the rabbits (Table 4 and Figure 2d), but the more time they spent in direct occupation with them (Table 4 and Figure 2b). Two models also revealed significant interactions with gender (Table 4): the decrease of time spent in proximity with age (Figure 2d) and the increase of time spent in direct occupation with age (Figure 2b) were both more pronounced in girls than in boys.

Effects of Social Interactions with Other Children and Child Personality

"Confident/Respected": The more "confident/respected" children (PCA factor 1; Table 3) spent proportionately more time in direct (Table 4 and Figure 3b) and indirect occupation with the rabbits (Table 4 and Figure 3c), and less time in proximity to them (tendency; Table 4 and

Table 4. Wald statistics for generalized linear models (GLMs) 1–4, with 1. "Frequency of occupation with rabbits" (mean number of entrances into observation range followed by engagement with rabbits), 2. "Duration of direct occupation with rabbits," 3. "Duration of indirect occupation with rabbits," and 4. "Duration of proximity to rabbits" (mean duration of categories [min/child/day]) as response variables.

Model Term	1: Frequency of Occupation with Rabbits Wald Statistic	df	p	2: Duration of Direct Occupation with Rabbits Wald Statistic	df	p	3: Duration of Indirect Occupation with Rabbits Wald Statistic	df	p	4: Duration of Proximity to Rabbits Wald Statistic	df	p
Gender	522.922	1	<0.001	39.233	1	<0.001	682.817	1	<0.001	125.389	1	<0.001
Siblings (yes/no)	167.926	1	<0.001	2048.388	1	<0.001	322.349	1	<0.001	47.788	1	<0.001
Pets (yes/no)	37.589	1	<0.001	436.365	1	<0.001	290.152	1	<0.001	341.513	1	<0.001
Age	322.581	1	<0.001	10.031	1	0.002	Excluded			552.346	1	<0.001
Confident/Respected	Excluded			163.977	1	<0.001	92.544	1	<0.001	3.352	1	0.067
Patient/Calm	247.900	1	<0.001	857.640	1	<0.001	169.487	1	<0.001	64.902	1	<0.001
Cheerful/Sociable	Excluded			121.046	1	<0.001	31.689	1	<0.001	11.749	1	0.001
Solitary	Excluded			39.671	1	<0.001	32.703	1	<0.001	143.415	1	<0.001
Gender * Siblings (yes/no)	21.494	1	<0.001	314.681	1	<0.001	Excluded			30.677	1	<0.001
Gender * Pets (yes/no)	45.407	1	<0.001	339.195	1	<0.001	4.329	1	0.037	136.028	1	<0.001
Gender * Age	Excluded			11.124	1	0.001	Excluded			55.430	1	<0.001
Gender * Confident/Respected	130.381	2	<0.001	18.646	1	<0.001	20.097	1	<0.001	136.091	1	<0.001
Gender * Patient/Calm	251.550	1	<0.001	1066.446	1	<0.001	Excluded			10.217	1	0.001
Gender * Cheerful/Sociable	26.634	2	<0.001	1044.577	1	<0.001	7.397	1	0.007	5.267	1	0.022
Gender * Solitary	36.524	2	<0.001	27.467	1	<0.001	Excluded			49.651	1	<0.001

Excluded = term was excluded from this final model because $p > 0.1$; * = interaction.

Table 5. Median frequencies (range) of occupation with rabbits (number of entrances into observation range followed by engagement with rabbits) per child per day and median duration (range) of occupation categories [min/child/day], per groupings of children according to sibling status and gender. Significant differences are shown in bold.

	Girls (n = 28)		Boys (n = 22)		Total (n = 50)	
Response Variable	W/ Siblings (n = 20)	W/O Siblings (n = 8)	W/ Siblings (n = 16)	W/O Siblings (n = 6)	W/ Siblings (n = 36)	W/O Siblings (n = 14)
Frequency of Occupation with Rabbits	2.53 (1.33–5.33)	2.63 (0.63–3.57)	1.44 (0.50–4.22)	1.44 (0.83–2.11)	2.06 (0.50–5.33)	1.47 (0.63–3.57)
Duration of Direct Occupation with Rabbits	4.81 (0.78–19.36)	2.37 (0.23–9.16)	2.36 (0.55–12.37)	0.52 (0.41–3.67)	3.37 (0.55–19.36)	1.27 (0.23–9.16)
Duration of Indirect Occupation with Rabbits	1.43 (0.19–4.63)	0.71 (0.02–2.27)	0.53 (0.02–2.00)	0.13 (0.02–1.50)	0.84 (0.02–4.63)	0.35 (0.02–2.27)
Duration of Proximity to Rabbits	2.08 (0.46–7.74)	1.86 (0.30–4.05)	1.26 (0.24–3.01)	1.49 (0.50–2.60)	1.32 (0.24–7.74)	1.49 (0.30–4.05)

Table 6. Median frequencies (range) of occupation with rabbits (number of entrances into observation range followed by engagement with rabbits) per child per day and median duration (range) of occupation categories [min/child/day], per groupings of children according to whether or not they had pets of their own and gender. Significant differences are shown in bold.

	Girls (n = 28)		Boys (n = 22)		Total (n = 50)	
Response Variable	W/ Pets (n = 16)	W/O Pets (n = 12)	W/ Pets (n = 6)	W/O Pets (n = 16)	W/ Pets (n = 22)	W/O Pets (n = 28)
Frequency of Occupation with Rabbits	2.22 (0.67–4.00)	3.06 (0.63–5.33)	1.55 (0.83–3.00)	1.44 (0.50–4.22)	2.00 (0.67–4.00)	1.95 (0.50–5.33)
Duration of Direct Occupation with Rabbits	2.70 (0.45–10.04)	4.81 (0.23–19.36)	2.36 (1.70–5.32)	1.43 (0.41–12.37)	2.55 (0.45–10.04)	3.06 (0.23–19.36)
Duration of Indirect Occupation with Rabbits	0.79 (0.02–3.19)	1.73 (0.02–4.63)	0.46 (0.14–0.87)	0.43 (0.02–2.00)	0.74 (0.02–3.19)	0.96 (0.02–4.63)
Duration of Proximity to Rabbits	0.89 (0.43–3.59)	3.14 (0.30–7.74)	1.00 (0.35–2.13)	1.34 (0.24–3.01)	0.89 (0.35–3.59)	1.66 (0.24–7.74)

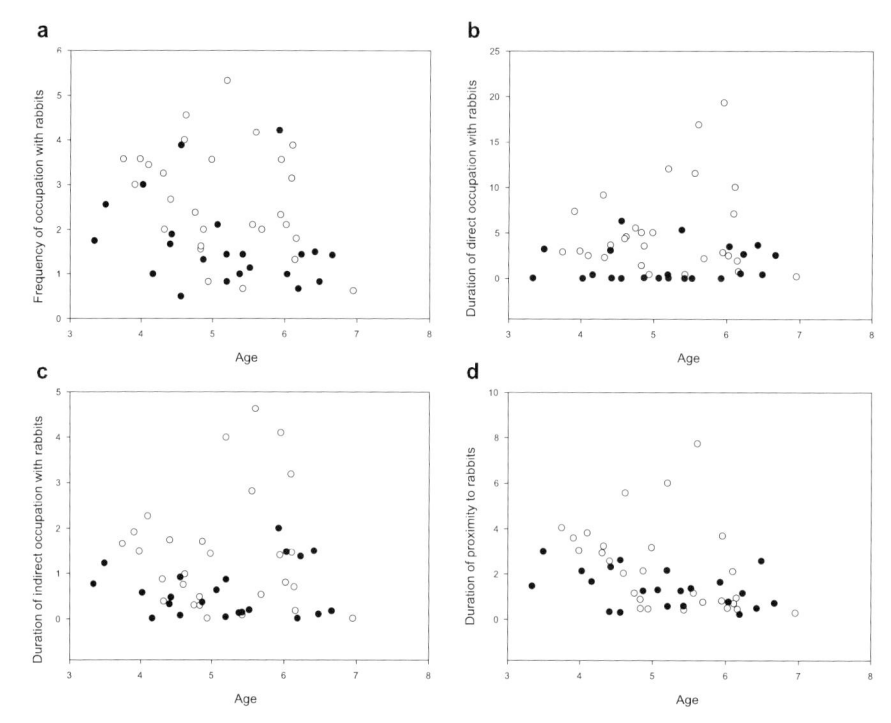

Figure 2. The relationship between age of child (years), and (a) mean frequency of occupation with rabbits (mean number of entrances into observation range followed by engagement with rabbits) per child per day and (b) mean duration of direct, (c) indirect occupation with rabbits, and (d) proximity to rabbits [min/child/day]; ○ girls (*n* = 28), ● boys (*n* = 22).

Figure 3d). All four models revealed significant interactions with gender (Table 4). The more "confident/respected" the boys, the less often they were occupied with the rabbits (Figure 3a), but the longer they were directly occupied with them (Figure 3b); this was not the case in girls. The more "confident/respected" the girls and boys, the longer they were indirectly occupied with the rabbits (Figure 3c), and the less time they spent in proximity to them (Figure 3d), but in boys these relationships were more pronounced.

"Patient/Calm": In general, the more "patient/calm" the children (PCA factor 2; Table 3), the less often they were occupied with the rabbits (Table 4) and the less time they spent with them (direct and indirect occupation and proximity to the rabbits; Table 4). However, three models revealed significant interactions with gender (Table 4). The more "patient/calm" the boys, the less often they were occupied with the rabbits and the less time they spent directly occupied with them, but in girls these relationships were not found. The more "patient/calm" the girls and boys, the less time they spent in proximity to the rabbits, but this relationship was more pronounced in boys.

"Cheerful/Sociable": The more "cheerful/sociable" the children (PCA factor 3; Table 3), the less time they spent in direct occupation with the rabbits (Table 4), but the more time they spent in indirect occupation with them (Table 4) and in proximity to them (Table 4). All four models revealed significant interactions with gender (Table 4). The more "cheerful/sociable" the boys,

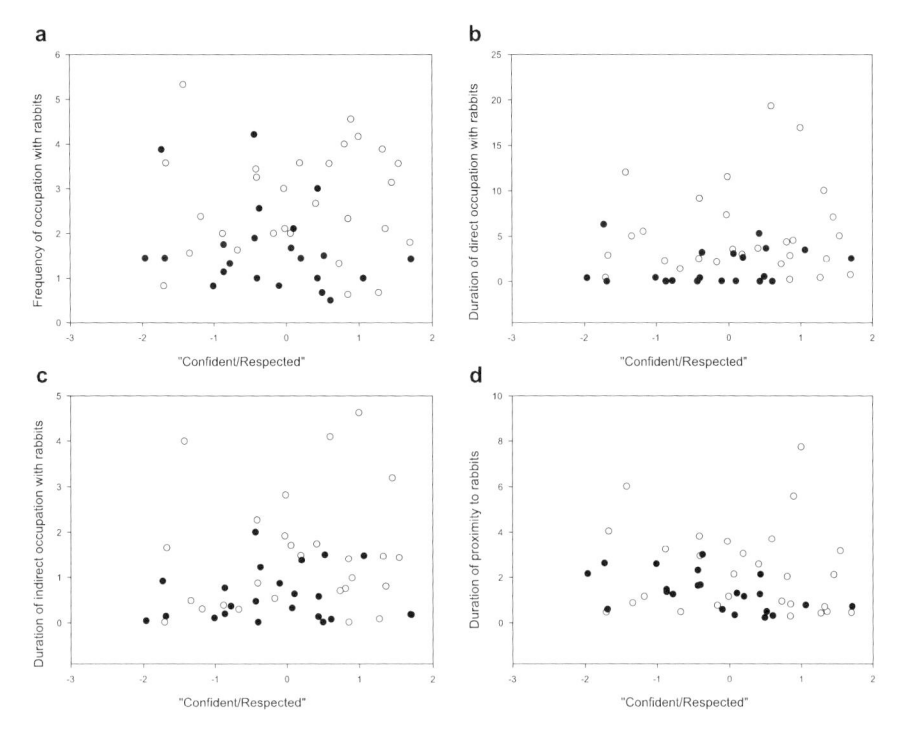

Figure 3. The relationship between the PCA factor 1 "Confident/Respected," and (a) mean frequency of occupation with rabbits (mean number of entrances into observation range followed by engagement with rabbits) per child per day and (b) mean duration of direct, (c) indirect occupation with rabbits, and (d) proximity to rabbits [min/child/day]; ○ girls (n = 28), ● boys (n = 22).

the less often they were occupied with the rabbits (vague relationship) and the less time they spent in direct occupation with them, but in girls there were no such relationships. The more "cheerful/sociable" the girls, the more time they spent in indirect occupation with the rabbits and in proximity to them, but in boys we did not find such relationships.

"Solitary": The more "solitary" (PCA factor 4; Table 3) the children, the less time they spent in direct and indirect occupation with the rabbits (Table 4), but the more time they spent in proximity to them (Table 4). Three models revealed significant interactions with gender (Table 4). The more "solitary" the boys and girls, the more often they were occupied with the rabbits, but in boys this relationship was more pronounced. The more "solitary" the boys, the more time they spent in direct occupation with the rabbits, while more "solitary" girls, spent proportionately less time in direct occupation with them; furthermore, the relationship in boys was more pronounced. The more "solitary" the boys, the more time they spent in proximity to the rabbits, but in girls we did not find this relationship.

Discussion

We found that the girls were generally more engaged with the rabbits than the boys. This result is consistent with those of a number of previous studies. For example, Paul and Serpell (1992) and Rost and Hartmann (1994) showed that girls were more likely to desire a pet than boys and maintained stronger emotional relationships with their pets (Ray 1982). Kotrschal and Ortbauer (2003) found no gender differences in interacting with a dog in a classroom of children aged 10 years, but behavior changes in the presence of the dog were more pronounced in the boys than in the girls.

Overall, our findings support the "social competence" hypothesis. In concurrence with findings that a strong bond between a child and a companion animal is contingent upon social competence and empathy (Poresky and Hendrix 1990; Poresky 1996), children high on the "confident/respected" and low on the "solitary" factors (Table 3) were particularly engaged in directly interacting with the rabbits. In contrast, less "confident/respected" and more "solitary" children spent greater amounts of time in proximity to, but less time directly occupied with, rabbits. The fact that these children, who often watched others play without joining in, also spent much time near the rabbits without directly or indirectly becoming occupied with them may indicate that they were shy or socially reluctant. However, they were not isolated or "socially incompetent." Perhaps some insecure, relatively low-ranking children are not especially interested in rabbits themselves, but benefit from them by getting into contact with other children in their area. Indeed, a "social lubricant effect" of animals has been reported by a number of researchers (e.g., Messent 1983; McNicholas and Collis 2000; Hergovich et al. 2002; Kotrschal and Ortbauer 2003). Animals seem to be excellent vehicles for conversation (Endenburg 2003).

Interestingly, children with pets of their own have been found to be more integrated in their class, embedded within more extended social networks, and generally more popular among their classmates (Endenburg and Baarda 1995) than children without pets. Taking care of a pet may indeed enhance self-confidence, social acceptance, and communication with humans (Messent 1983; Hunt, Hart and Gomulkiewicz 1992; Poresky 1996; McNicholas and Collis 2000; Endenburg 2003). Alternatively, it may well be that families that allow pet-keeping may provide a different social environment than families that don't. In the present study, children without pets were significantly more often occupied with the rabbits and spent significantly greater amounts of time directly and indirectly engaged with, and in proximity to, them than pet-owning children. Children with siblings were generally more intensely occupied with the rabbits than children without siblings (frequency of occupation with rabbits and duration of direct and indirect occupation with rabbits and staying in proximity to rabbits). These differences parallel previous findings that family composition affects attitudes towards animals (Godwin 1975; Franti et al. 1980; Melson 1988; Kidd and Kidd 1989; Endenburg, Hart and de Vries 1990). Generally, in the present study, pet-owning and having siblings had significant effects on frequency and duration of occupation with rabbits. In girls, both pet-owning and having siblings had significant effects on frequency and duration of occupation with rabbits. In boys, however, the intensity of direct occupation with the rabbits was not related to having a pet at home but was by having siblings.

We investigated whether and how differential demonstration of interest in animals could be related to age, gender, family background, play behavior, and personality components—we considered these variables to be the most important factors in children of this age group. We did not investigate the influence of other factors such as self-esteem, social anxiety, stress level, emotional competence, self-concept, or peer group influences, primarily because of the low sample size. These factors would be interesting to address in future studies and are necessary to further our understanding of interest in contact and occupation with animals. Also, mainly because of the low sample size, we could not include exploration of pet personality features or more information about family background (e.g., age and gender of siblings; kinds of, and relationship to, pets at home; and parent's attitudes towards pets). Also, in future studies other pet animals should be considered.

Because in this preschool many children passed through the observation areas for reasons unrelated to interest in the rabbits, we included only those children who were occupied at least once with the animals while in the observation field. To investigate the idea that socially "incompetent" children may spend time near the animals without being occupied with them at all, a study with more controlled conditions should be conducted, in which the animals are kept in an area that children would approach only to spend time around them or be occupied with them.

Longitudinal studies would provide a better understanding of the role of animals in different stages of a child's development. Because starting preschool constitutes an important change in a child's life, a study focusing on the first year of attendance would provide opportunity to examine the potential influence of interest in animals on children's stress levels and the development of, and changes in, friendships and social structure.

In conclusion, our findings indicate that girls are more interested in rabbits than are boys. Children low in social status and children who watch others play rather than playing actively with others themselves tended to linger in proximity to the rabbits, rather than interacting actively with them. In contrast, the high-status, socially interactive children actively interacted with the rabbits. Hence, the social styles of these children were reflected in the way they interacted with the animals.

Acknowledgements

Financial and logistic support was provided by IEMT Austria. We thank the preschool teachers and assistants of the *Erlebniskindergarten-Krems-Mitterau I*, particularly Annemarie Donabaum, the director of the preschool, for their generous help and cooperation. We would also particularly like to thank the children and parents whose participation made this study possible. This study could not have been conducted without general interest on the part of the Austrian governmental agencies that granted permits: *Bezirkshauptmannschaft Krems, Magistrat der Stadt Krems and Abteilung Kindergärten des Amtes der Niederösterreichischen Landesregierung*. Our thanks also go to two anonymous referees of this paper who provided valuable and constructive comments on the manuscript. Finally, we are grateful to Barbara Bauer, Dorothy Gracey, Daniela Lexer, Elisabeth Oberzaucher, Isabella Scheiber, Claudia Schmied, Susanne Waiblinger and Brigitte Weiß for their assistance and support.

References

Bachmann, R. W. 1975. Elementary school children perception of helpers and their characteristics. *Elementary School Guidance and Counseling* 10: 103–109.

Brickel, C. M. 1982. Pet facilitated psychotherapy. A theoretical explanation via attention shifts. *Psychological Reports* 50: 71–74.

Collis, G. M. and McNicholas, J. 1998. A theoretical basis for health benefits of pet ownership: Attachment versus psychological support. In *Companion Animals in Human Health*, 105–122, ed. C. C. Wilson and D. C. Turner. Thousand Oaks, CA: Sage Publications.

Endenburg, N. 2003. Tiere in der Entwicklung und Psychotherapie: Der Einfluss von Tieren auf die Frühentwicklung von Kindern als Voraussetzung für tiergestützte Psychotherapie. In *Menschen brauchen Tiere: Grundlagen und Praxis der tiergestützten Pädagogik und Therapie*, 121–130, ed. E. Olbrich and C. Otterstedt. Stuttgart: Franckh-Kosmos.

Endenburg, N. and Baarda, B. 1995. The role of pets in enhancing human well-being: Effects on child development. In *The Waltham Book of Human–Animal Interaction: Benefits and Responsibilities of Pet Ownership*, 7–17, ed. I. H. Robinson. Exeter: Pergamon.

Endenburg, N., Hart, H. and de Vries, H. W. 1990. Differences between owners and non-owners of companion animals. *Anthrozoös* 4: 120–127.

Franti, C. E., Kraus, J. F., Borhani, N. O., Johnson, S. L. and Tucker, S. D. 1980. Pet ownership in rural northern California (El Dorado County). *Journal of the American Veterinary Medical Association* 176: 143–209.

Friedmann, E. 2000. The animal–human bond: Health and wellness. In *Handbook on Animal-Assisted Therapy: Theoretical Foundations and Guidelines for Practice,* 41–58, ed. A. H. Fine. San Diego: Academic Press.

Godwin, R. D. 1975. Trends in the ownership of domestic pets in Great Britain. In *Pet Animals and Society,* 96–102, ed. R. S. Anderson. London: Baillière Tindall.

Grammer, K. 1995. *Biologische Grundlagen des Sozialverhaltens: Verhaltensforschung in Kindergruppen.* Darmstadt: Wissenschaftliche Buchgesellschaft.

Guttman, G., Predovic, M. and Zemanek, M. 1985. The influence of pet ownership in non-verbal communication and social competence in children. In *The Human–Pet Relationship: Proceeding of the International Symposium on the Occasion of the 80th Birthday of Nobel Prize Winner Prof. DDr. Konrad Lorenz, October 27–28, 1983,* 58–63. Vienna: IEMT.

Hergovich, A., Monshi, B., Semmler, G. and Ziegler, V. 2002. The effects of the presence of a dog in the classroom. *Anthrozoös* 15: 37–50.

Herzog, H. A. 2007. Gender differences in human–animal interactions: A review. *Anthrozoös* 20: 7–21.

Hold, B. 1976. *Rangordnungsverhalten bei Vorschulkindern. Homo* 25: 252–267.

Hunt, S. J., Hart, L. A. and Gomulkiewicz, R. 1992. Role of small animals in social interactions between strangers. *The Journal of Social Psychology* 132: 245–256.

Katcher, A. H. and Beck, A. M. 1985. Sicherheit und Vertrautheit. Physiologische und Verhaltensreaktionen auf die Interaktion mit Haustieren. In *Die Mensch-Tier-Beziehung: Dokumentation des Internationalen Symposiums aus Anlass des 80. Geburtstages von Nobelpreisträgers Prof. DDr. Konrad Lorenz, 27. und 28. Oktober 1983,* 131–138. Wien: IEMT.

Kidd, A. H. and Kidd, R. M. 1989. Factors in adults' attitudes towards pets. *Psychological Reports* 65: 903–910.

Kotrschal, K. and Ortbauer, B. 2003. Behavioral effects of the presence of a dog in a classroom. *Anthrozoös* 16: 147–159.

McNicholas, J. and Collis, G. 2000. Dogs as catalysts for social interactions: Robustness of the effect. *British Journal of Psychology* 91: 61–70.

Melson, G. F. 1988. Availability and involvement with pets by children: Determinants and correlates. *Anthrozoös* 2: 45–52.

Messent, P. R. 1983. Facilitation social interaction by companion animals. In *New Perspectives on Our Lives with Companion Animals,* 37–46, ed. A. H. Katcher and A. M. Beck. Philadelphia: University of Pennsylvania Press.

Mugford, R. A. and M'Comisky, J. G. M. 1975. Some recent work on the psychotherapeutic value of cage birds with old people. In *Pet Animals and Society,* 54–65, ed. R. S. Anderson. London: Baillière Tindall.

Paul, E. S. 1995. Us and them: Scientists' and animal rights campaigners' views of the animal experimentation debate. *Society & Animals* 3: 1–21.

Paul, E. S. 2000. Love of pets and love of people. In *Companion Animals & Us: Exploring the Relationships between People and Pets,* 168–186, ed. A. L. Podberscek, E. S. Paul and J. A. Serpell. Cambridge: Cambridge University Press.

Paul, E. S. and Serpell, J. 1992. Why children keep pets: the influence of child and family characteristics. *Anthrozoös* 5: 231–244.

Poesel, A., Kunc, H. P., Foerster, K. and Kempenaers, B. 2006. Early birds are sexy: Male age, dawn song and extrapair paternity in blue tits, *Cyanistes* (formerly *Parus*) *caeruleus. Animal Behaviour* 72: 531–538.

Poresky, R. H. 1996. Companion animals and other factors affecting young children's development. *Anthrozoös* 9: 159–168.

Poresky, R. H. and Hendrix, C. 1990. Differential effects of pet presence and pet bonding on young children. *Psychological Reports* 67: 51–54.

Ray, J. J. 1982. Love of animals and love of people. *Journal of Social Psychology* 116: 299–300.

Rost, D. H. and Hartmann, A. 1994. Children and their pets. *Anthrozoös* 7: 242–254.

Seabrook, M. F. 1984. The psychological interaction between the stockman and his animals and its influence on performance of pigs and dairy cows. *The Veterinary Record* 115: 84–87.

Serpell, J. 1986. *In the Company of Animals: A Study of Human–Animal Relationships.* Oxford: Basil Blackwell.

Waiblinger, S. 1996. Die Mensch-Tier-Beziehung bei der Laufstallhaltung von behornten Milchkühen. *Tierhaltung* 24, Ökologie, Ethologie, Gesundheit, Universität/Gesamthochschule Kassel.

Wilson, E. O. 1984. *Biophilia.* Cambridge, MA: Harvard University Press.

NEWS & ANALYSIS

Journal of Social Issues— Special Issue

New Perspectives on Human–Animal Interactions: Theory, Policy, and Research

This special issue (2009, Volume 65, Number 3) of the *Journal of Social Issues* contains ten articles by established researchers on psychological aspects of our interactions with members of other species. Edited by Sarah Knight and Harold Herzog, the issue is divided into three sections: Factors underlying attitudes and behaviors toward the treatment of other species; the effects of animals on human health and well being; and the roles of animals in human society. Topics include between-group differences in attitudes toward animal use, the role of moral emotions in animal activism, ethical implications of animal consciousness, the effects of companion animals on human physical and psychological health, interactions with robotic pets, the relationship between animal cruelty and human violence, and the consumption of dogs and cats in Asia. In the final chapter, James Serpell explains why animals are an important social issue.

New Books

What Species of Creatures: Animal Relations from the New World

By Sharon Kirsch

This book chronicles how the first European arrivals to Canada experienced wild animals previously unknown to them: the flying squirrel, the ruby-throated hummingbird, the white or polar bear. Drawing on, and refashioning traditional genres of, animal writing (the fable, children's stories, classifications by naturalists, and even merchandise lists from the Hudson's Bay Company), *What Species of Creatures* explores the origins of our present-day attitudes toward animals, gently challenging readers to consider their own place in the hierarchy of "beasts." Published in 2008 by New Star Books. ISBN-13: 978-1-55420-040-5 (paperback).

Kinship and Killing: The Animal in World Religions

By Katherine Wills Perlo

Through close readings of Jewish, Christian, Islamic, and Buddhist texts, Katherine Wills Perlo proves that our relationships with animals shape religious doctrine, particularly through the tension between animal exploitation and the bonds of kinship. And as humans feel more empathy toward animals, Perlo finds that adherents revise their interpretations of religious texts. Preexisting ontologies, such as Christianity's changing God or Buddhism's principle of impermanence, along with advances in farming practices and technology, also encourage changes in treatment. As cultures begin to appreciate the different types of perception and consciousness experienced by nonhumans, definitions of reality become complicated and humans lean more toward unitary accounts of shared existence. These evolving attitudes exert a crucial influence on religious thought, Perlo argues, moving humans ever closer to a nonspeciesist world. Published in 2009 by Columbia University Press. ISBN: 978-0-231-14623-4 (paperback); ISBN: 978-0-231-14622-7 (hardback).

Noble Brutes: How Eastern Horses Transformed English Culture

By Donna Landry

More than 200 Oriental horses were imported into the British Isles between 1650 and 1750. With the horses came Eastern ideas about horsemanship and the relationship between horses and humans. Landry's groundbreaking archival research reveals how these Eastern imports profoundly influenced riding and racing styles, as well as literature and sporting art. Published in 2009 by The Johns Hopkins University Press. ISBN: 978-0-8018-9028-4 (hardback).

Animal Encounters

Edited by Tom Tyler and Manuela Rossini

Animal Encounters comprises a series of meetings not only between diverse beasts, but also between distinct disciplinary methods, theoretical approaches, and ethical positions. The essays here collected come together from literary and cultural studies, sociology and anthropology, ecocriticism and art history, philosophy and feminism, science and technology studies, history, and posthumanism, to study that most familiar and most foreign of creatures, "the animal." Published in 2009 by Brill. ISBN-13: 978-90-04-16867-1 (paperback).

Animal Capital: Rendering Life in Biopolitical Times

By Nicole Shukin

The juxtaposition of biopolitical critique and animal studies—two subjects seldom theorized together—signals the double-edged intervention of *Animal Capital*. Nicole Shukin pursues a resolutely materialist engagement with the "question of the animal," challenging the philosophical idealism that has dogged the question by tracing how the politics of capital and of animal life impinge on one another in market cultures of the twentieth and early twenty-first centuries. Published in 2009 by University of Minnesota Press. ISBN: 978-0-8166-5342-3 (paperback).

Conferences

19th Annual ISAZ Conference
Cross-Cultural Attitudes Toward Animals

June 30, 2010

The 19th Annual ISAZ Conference, being held in conjunction with the 12th Meeting of the International Association of Human–Animal Interaction Organizations (IAHAIO), will be held in

Stockholm, Sweden and will focus on the important and emerging body of research on cross-cultural attitudes toward animals. There will be space in the program for papers outside the conference theme, and information related to abstract submission will be forthcoming. Additional information will be circulated and posted on the ISAZ website, as it becomes available: www.isaz.net. Information on the conference location can be viewed at: www.stoccc.se/en/folkets-hus/folkets-hus/. If you have any questions, please send an e-mail to: isaz2010@isaz.net

12th IAHAIO Conference
People and Animals: For Life

July 1 to 4, 2010

This conference of the International Association of Human–Animal Interaction Organizations will be held in **Stockholm, Sweden**. The goals of the conference are:

1. To present the latest research findings in the area of human–animal interactions, including animals in therapeutic interventions, as social and educational support, and as an integral part of a health-providing society.

2. To provide a forum for international networking amongst practitioners and researchers in the various fields of human–animal interaction.

3. To focus especially on the use of horses in therapy, and on the impact of animals on human function, activity, and participation in society.

Target audience: includes researchers and practitioners in the field of human–animal interactions, including medical, veterinary, behavioral, educational, and social sciences, as well as those involved in planning and developing a sustainable society. Students of the above-mentioned disciplines and professionals in animal-related fields such as service dog organizations and their clients, zoo professionals, animal welfare organizations, various government bodies, animal trainers and groomers, and other pet-related businesses are also welcome. In addition, researchers and students of social planning, environmental and rural development, health promotion, sports, and healthy lifestyles for humans are special target groups.

For further information, send an e-mail to: iahaio2010@congrex.com, or visit the official web site: www.iahaio2010.com.

11th International Council for Archaeozoology (ICAZ) Conference

August 23 to 28, 2010

This conference will take place in **Paris, France.** Some events will be organized at the Muséum National d'Histoire Naturelle, but most of the conference will take place at the nearby campus of Pierre et Marie Curie University (Jussieu). For further information, please send an e-mail to: icaz2010@mnhn.fr

At Rest with the Animals
THOUGHTS OVER THIRTY YEARS
Colman McCarthy
The Humane Society of the United States

discover a new favorite

At Rest with the Animals, Colman McCarthy
ISBN 0-974-8400-4-1, *$21.95, Paper, 8.625" x 8.5"*

Order online at **humanesociety.org/hspress**
Mail orders to Humane Society Press
2100 L Street, NW, Washington, DC 20037

Add $3.00 shipping/handling per order

THE HUMANE SOCIETY OF THE UNITED STATES
Celebrating Animals | Confronting Cruelty

BOOK REVIEWS

Swan

Peter Young. London: Reaktion Books, 2008. 200 pages. ISBN: 978-1-86189-349-9 (paperback)

Reviewed by: Patricia K. Anderson, Department of Sociology & Anthropology, Western Illinois University, USA. E-mail: PK-Anderson@wiu.edu

Swan, by Peter Young, who also authored *Tortoise*, is another fine contribution to the Reaktion *Animal* book series. This book is beautifully illustrated and replete with fascinating details of the natural history of swans, in addition to rich historical and cultural facts regarding the nominate animal.

The book is organized into 9 chapters. The first, "Clamorous Wings," provides an introduction to the birds with discussion of ancestry, reproduction, life-span, extant species, and geographic distribution, migration, and other details including the various terms used by humans to designate swans, male (cob) from female (pen), and their young (cygnets). Collective nouns for groupings of swans include "bank," "bevy" (also used for quails and larks), "a drift," "a game," "a herd," "a lamentation," "a sownder," "a whiting," or "team" (pp. 17–19). An interesting note is the relatively long life span—potentially up to 50 years—of captive swans. I found this particularly interesting since the boundary between human and swan is ephemeral in folklore, an interesting coincidence that the author might have explored.

In Chapter 2, "Grace and Favour," the author investigates a rich heritage of swan symbolism in literature, art, music, and metaphor. Here the great range of influence of the concept swan in human thought becomes clear. Examples from myth (also considered in Chapter 3 including the obvious reference to Leda), to poetry to song to sculpture, to whimsical automobile, the "unique Swan Car" (p. 48), to the swan yoga position (the hamsamsana), the influence of the notion of "swanness" is wide-felt. Swans have long been associated with death and prophecy as in the saying, "swan song," relating to the animal's alleged ability to presage its own death through song. We also find that the very social bird will become depressed in isolation and that in recovering from surgery, swans benefit from the social support of their own kind. The legend that swans mate for life and die of grief is apparently untrue, however (p. 43).

Chapter 3, "Transformations," deals with the metamorphic symbolic nature of the swan in folklore, literature and art. Cross-culturally, an interesting commonality is the legend of the swan maiden, a dual natured being who can transform from human to swan with the aid of a magical robe of swan feathers or the reverse with its removal. It is interesting that the shape shifter continues to be part of modern human thought and culture through such popular genres as literature and film in the popular Twilight series and the "furry" subculture. Why do these boundary creatures continue to be important in human thought? And more importantly, what do the swan maiden myths tell us about human attitudes toward swans? The author might have profitably explored these questions in this chapter, although this is perhaps best addressed elsewhere due to the brevity of the book.

The fourth chapter, "Swans in History," indicates that the Image or name of the bird has left an indelible mark on human cultures, with its name and image linked to place names, inns, taverns, consumer products, and family coats of arms, among many others.

The most important threats to swans today are legal and illegal human activities (Chapter 5). Hunting of tundra swans is still allowed in some states in the USA. Although swans have largely positive associations in human thought and culture, that does not protect the birds from those who seek them out for target practice and or torture, resulting in the killing or maiming of the defenseless birds. Fish lines and fishhooks are another common threat and photographs of injured swans with fishhooks in their necks appear (pp. 122–123). Without veterinary intervention the birds are likely to slowly starve to death if they don't die first from the wounds. Death from the ingestion of lead fishing weights is also a major concern. Although legislation in Great Britain has forced law-abiding anglers to switch to less toxic materials, the birds are still exposed, especially during times of drought when water levels fall, thus exposing residual lead weights in the silt where the birds forage. Another source of lead poisoning is residual gunshot that the birds may innocently ingest as grit to aid in digestion. Threats posed by anthropogenic structures such as power lines, wind turbines, and glass windows, are also discussed, as well as a potential solution—locate future constructions outside the paths of flight patterns. Threats to both people and swans potentially occur when human air traffic crosses the path of swans in flight, and swans may collide with airplanes or be sucked into jet engines, resulting in the loss of human and swan lives as well as aircraft.

Consumption of swan byproducts is also discussed in this chapter, especially the production and consumption of swan feather quills, which are reputed to be the highest quality and much preferred by scribes. For over a thousand years, quills were the main instruments used in writing. Swan quill pens are still used today for ceremonial and other special occasions, but feather quill pens were largely replaced during the 1820s with the mass production of the pressed steel nib. Swan feathers have also historically been used to trim hats, fans, muffs, and make other articles of clothing and sundry items. The down, for example, was used to stuff mattress quilts, quilted clothing, and to make cosmetic powder puffs for the elite. It is interesting that swan feathers are still used to trim ceremonial attire for the Queen's bodyguard, and other dignitaries. Use of swan feathers by tribal societies is given brief coverage, with mention of the perceived spiritual power of swan feathers used by shamans in healing (p. 135).

Historically, it is not surprising that the large bird has long been used as food, the topic of Chapter 6. Although classified as unclean meat in the Old Testament, swan meat was consumed by wealthy Europeans from the eighth to eighteenth centuries, the practice only fading in popularity with the introduction of the turkey from the New World. When served at court (presumably in England, although it is not specified), swans were skinned before cooking and served whole, with its wings wired in an animate position as if about to fly.

The topic of Chapter 7 (pp. 150–167) is conservation, although conservation is not mentioned until the reader reaches the latter part of page 153. Instead, the chapter commences with the description of the system of marking that was historically used to designate ownership of individual birds. In England all unmarked swans were considered property of the Crown, but others were distinguished by a pattern of notches carved in the bill. These notch patterns were registered and considered property that could be passed down to one's heirs, a practice that dates back to at least 966 when King Edgar gave rights over straying local mute swans to the Abbots of Croyland. Penalties for poaching swans or stealing their eggs were severe, and during the mid-nineteenth century included deportation to the penal colony in Australia. The annual practice of swan "upping" or swan "hopping," begun on the Thames River in the twelfth century, involved the marking and census of swans. The practice continues today with colorful pageantry held in late July, although swan marking has been replaced by banding. Modern responsibilities of the Queen's Swan Marker include removal of the birds during rowing events on the Thames and education on swan conservation and welfare. Conservation efforts and status of the bird in North and South America and Asia are also discussed. Currently, swan populations are returning after a precipitous dip during the early part of the last century due to the commercial hunting for meat and feathers, as well as the loss of habitat. The trumpeter swan, the largest of the North American swans, was nearly decimated. Reintroduction efforts have been successful, however, and the survival of the species now seems certain.

Not all people are fans of swans, however, and the introduced mute swan is perceived as a threat to native species in North America but championed by animal rights advocates. In England, fisherman consider the birds destructive to the aquatic vegetation upon which fish feed, even though pesticides and changes in water use may be the real culprits.

Chapter 8, entitled "Promotion," discusses the wide usage of the bird in advertising. The image and name have been used to hawk countless items, from the very popular Vesta matches and cigarettes in the United Kingdom, to fashions, cosmetics, football teams, glassware, horse brasses, software companies, and more. Consumers are encouraged to look for the swan logo, the official Nordic "ecolabel," found on environmentally friendly products in Finland, Norway, and Sweden.

Surprisingly, the term "swan," was used during the 1990s as slang for the illicit drug Ecstasy, deriving from the image stamped on some of the tablets (p. 180).

Threats to swans might have been more profitably combined with the final chapter on conservation, while the section discussing swan byproducts could have been combined with the "food" chapter under the topic "consumption." Further, ornithologists and others who study swans may find the work lacking with omission of reference to Banko's classic study, *The Trumpeter Swan*, as well as A. Lindsey Price's *Swans of the World, in Nature, History, Myth, and Art*. However, it is difficult to provide a comprehensive overview in 200 pages, and these are minor objections. Therefore, *Swan* quite nicely meets the goal of the Reaktion book series, which is to focus on the impact of the title animal on human history and culture.

The International Handbook of Animal Abuse and Cruelty: Theory, Research and Application

Edited by Frank Ascione. West Lafayette, IN: Purdue University Press, 2008. 520 pages. ISBN-13: 978-1-55753-463-7 (hardback)

Reviewed by: Alison Reynolds, Training, Support & Research Manager, Pets As Therapy, UK. E-mail: areynolds@petsastherapy.org

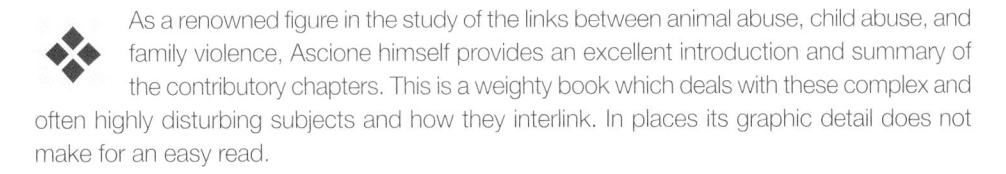

As a renowned figure in the study of the links between animal abuse, child abuse, and family violence, Ascione himself provides an excellent introduction and summary of the contributory chapters. This is a weighty book which deals with these complex and often highly disturbing subjects and how they interlink. In places its graphic detail does not make for an easy read.

Whilst there is an unarguably impressive authorship, the contributors are largely American and this is reflected in the majority of the content throughout the handbook. There is an international flavor in parts, with interesting chapters by Yokoyama and by Maruyama and Ascione on the unique cultural and societal approaches to animal welfare in Japan, and by Pagani and others in Italy, but the majority of the content is USA-focused. This may limit its applicability to other countries, particularly in relation to the legislative aspects.

The predominant human healthcare field from which contributions are drawn is psychology, with other input from psychiatry, social work, and sociology. Inputs from a legislative perspective are provided by Frasch and Schuleter, with other authors referring to legal implications or applications throughout. Somewhat surprisingly, given the book's subject, contributions from animal care and welfare advocates, for example, veterinarians, animal behaviorists, and animal psychologists, form only a minority of the contributors.

Many of the chapters provide in-depth theoretical perspectives, with well-researched and referenced evidence. Unti's chapter begins by documenting the progress in animal protection in America since the 16th Century and explains the inevitable link with child protection services, contextualized within other social and political contexts, for example, feminism. He explains how, paradoxically, animal protection as a social reform was lost with modern (20th Century) ideas of anthropocentrism, scientific and technical advances, and increasing diversity in humans' use of animals, which brings the reader up to the current-day situation.

Chapter 2, co-authored by Arkow and UK veterinary surgeon Munro, is an important chapter because there are only two veterinary contributors to this handbook. The authors present the difficulties which face and largely prevent the veterinary profession from recognizing and preventing family violence, including: inconsistent definitions, the need for client confidentiality, and the fear of litigation. They include the survey results for the prevalence of non-accidental injury (NAI) experienced by veterinarians from several different countries, which gives the chapter international applicability, and offer case studies and diagnostic indicators for NAI. They conclude with concise recommendations for a developmental agenda.

Frasch examines the impact of improved anti-cruelty laws on prosecutions for animal abuse in the USA. This is an informative chapter, providing state-by-state summaries, but may be somewhat limited in its interpretation and application by being solely USA-focused.

Several authors refer to attempts to measure or assess the prevalence of animal abuse, child abuse, and domestic violence in particular countries, states, or regions. Lockwood's chapter on "counting cruelty," although again American-focused, gives a broad overview of the rationale for trying to assess different forms of violence, the type of information that needs to be gathered, and some solutions for applying this to increase our knowledge of the scale of the problem.

As a clinical psychologist, Dadds presents contemporary models which outline the possible causes of conduct problems in children and their relationship to animal cruelty within a cluster of both violent and non-violent forms of anti-social behaviors. This is the first mention of the "Triad" of behaviors of fire-setting, enuresis, and animal cruelty, and the evidence is reviewed and presented as tenuous. Dadds suggests potential strategies for identifying animal cruelty, and although he suggests that the veterinary profession plays a key role, no strategies for their involvement are suggested here.

Magid, too, explores the relationships between attachment theory and animal abuse, which he illustrates through two opposing case examples. The chapter is divided into four parts: the first deals with the fundamentals of attachment theory; the second with attachment pathways as risk factors for clinical problems, and the third with the links between the degree of attachment disorder and the likelihood of perpetrating animal abuse. Part four explores our evolutionary connections with animals and the challenges for future generations.

Maiuro, Eberle, Rastaman and Snowflake bring a joint perspective from animal and human psychology to attachment issues and their relationship with animal abuse. The accessible writing style makes a complex subject easy to digest. The possible links between psychobiology, attachment theory, domestic violence, and animal abuse are explored with reference to some interesting historical and contemporary research studies. The chapter concludes with a set of clear and helpful recommendations and discussion points.

Flynn argues for a sociological analysis of animal abuse and cruelty, rather than a search for the characteristics of individual perpetrators, and he discusses issues such as gender, age, cultural attitudes, and power and inequality. He includes a number of theoretical approaches, for example, feminism and symbolism, and then examines a number of studies on animal abuse data. He is one of only two contributors to specifically mention "The Link," but does not provide a definition of it; instead, he discusses difficulties with its acceptance, particularly in relation to the methods and studies on which evidence for it is based.

Faver and Strand's chapter offers a viewpoint on the role that social workers can play in combating animal abuse. They provide a clear rationale for the involvement of social work and offer a number of practical strategies related to community participation and education, screening and assessment tools, and other practical resources, such as videos, websites, and books.

In Beetz's chapter on bestiality, she provides an overview of the existing knowledge base on this taboo subject, explores the historical evidence, particularly in relation to legislation, and gives an indication of the prevalence and species involvement since 1948. The issue of inconsistency in definitions is highlighted and, within this context, Beetz asks the reader to consider the issue of animal consent, or perceived consent within human–animal sexual relationships. There is a very brief mention of animal welfare and animal dignity and the relevance of the study of bestiality to veterinary science.

Patronek's chapter provides an interesting insight into the poorly understood behavior of animal hoarding. Whilst animal abuse/cruelty arise through the failure to provide adequate care, the underlying motivations and intentions are quite different and often occur in the presence of a strong and positive human–animal bond. Some treatment and intervention models are briefly considered and again an assessment form for social work use.

There are several chapters offering specific cultural perspectives: from Japan, by Yokoyama, and Maruyama and Ascione; from Italy, by Pagani, Robustelli and Ascione; and an Australian perspective is offered by Gullone and Clark.

Pagani, Robustelli and Ascione's chapter is based on their study of 800 Italian schoolchildren's experiences of/exposure to animal abuse and cruelty, including motivational aspects of committing cruel acts and perceptions of socially acceptable and unacceptable forms of violence towards animals. The chapter provides an interesting description of Italian cultural beliefs and attitudes towards animal welfare and the lack of priority assigned to it socially. They also discuss more recent positive and negative societal changes, such as increased pet ownership and increases in ritualistic animal sacrifice and dog-fighting.

Maruyama and Ascione provide an interesting insight into Japanese culture and the emergence of knowledge about the extent of animal abuse and cruelty, particularly in relation to the main social problems unique to this country. Japan is described as culturally isolated and strongly influenced by gender-appropriate behaviors (a historical value which persists and is encouraged), both of which have implications for the incidence and prevalence of animal abuse and cruelty.

They describe a "social crisis" in Japan, with a massive increase in violent juvenile crime, believed to be strongly linked to wealth and to the extremely high expectations placed upon young people in Japanese society; child abuse and abuse of elders, too, are on the increase. The first half of the chapter is largely descriptive and the omission of any supporting evidence does undermine some of the suggested links. The second half of the chapter deals with the risk factors for animal abuse and the significant social issues faced by a modern Japanese society, which have only recently been recognized. The authors recognize the difficulties of cross-cultural comparisons and interventions and also highlight the lack of research in Japan.

An Australian perspective on animal abuse and cruelty is offered by Gullone and Clarke and begins with a current review of Australian attitudes and beliefs towards animal cruelty, as well as legislative perspectives. They highlight the substantial variations in penalties enforced in different states and territories, and provide data on companion animal ownership, dog bites and attacks, and statistics and trends on animal abandonment, relinquishment, and abuse. They then review some of the evidence for explaining animal abuse and its link with other forms of family violence and criminal behavior, and the remainder of the chapter presents a series of tables from Victoria with data on animal abuse offences in relation to other types of offences.

Schlueter provides a largely personal perspective on American law enforcement perspectives and obligations which hinder the study and prevention of animal abuse and cruelty, and briefly offers some recommendations. Despite the personal perspective, the recommendations are broad enough to be appropriately applied in any country.

The chapter by Boat, Loar and Phillips provides a rationale for cross-collaboration between mental health, child protection, and prosecution officials and is probably the only chapter which specifically sets out to offer a practical strategy for cross-collaboration. After presenting the rationale for involvement of each of the three parties, the authors include case examples and copies of assessment tools.

In their focus on children's experiences with animal cruelty, through reference to various key research studies, Lewchanin and Randour suggest intervention strategies which could be broadly applied elsewhere, although the remainder of the chapter is again only applicable to the USA.

Yokoyama brings a further Japanese perspective to the final chapter, where he discusses psychiatric disorders, for example, schizophrenia, mood and anxiety disorders, and dementia, within the context of both positive and negative human–animal interactions. He proposes a clarification of animal abuse and categorization into three areas, based on a research study of 862 veterinary hospitals. In agreement with Maruyama and Ascione, he acknowledges that the study of animal abuse is in its infancy in Japan, but also attempts some generalizations to apply to other countries.

Ascione has successfully achieved the task he was set—"to assemble a compendium of papers reflecting the current state of scholarly work." The book cannot be faulted in its extensive coverage of relevant and current theory and research, and is hugely informative. Whilst many of the contributors do provide some useful practical tools, for example, assessment, screening, and interview schedules, practical strategies for implementation of multi-disciplinary programs and cross-reporting may be more limited. There are many excellent recommendations made for future research agendas and there are references made to other types of human–animal interactions of a more positive nature, namely Animal-Assisted Therapy and Animal-Assisted Activities, both of which could be used to prevent animal abuse or rehabilitate high-risk groups. The book falls more within the realms of a valuable and informative scholarly resource than a practical handbook, and so may have wider appeal to a more academic audience.

INDEX

LIST OF ARTICLES

Volume 22, Number 1

Avian Consciousness in Don DeLillo's The Body Artist
Jeffrey Karnicky — 5

Vampires Are Still Alive: Slovakian Students' Attitudes toward Bats
Pavol Prokop, Jana Fančovičová and Milan Kubiatko — 19

An Examination of Changes in Oxytocin Levels in Men and Women Before and After Interaction with a Bonded Dog
Suzanne C. Miller, Cathy Kennedy, Dale DeVoe, Matthew Hickey, Tracy Nelson and Lori Kogan — 31

Comparison of the Effect of Human Interaction, Animal-Assisted Therapy, and AIBO-Assisted Therapy on Long-Term Care Residents with Dementia
Stephen C. Kramer, Erika Friedmann and Penny L. Bernstein — 43

Exploring Stock Managers' Perceptions of the Human–Animal Relationship on Dairy Farms and an Association with Milk Production
Catherine Bertenshaw and Peter Rowlinson — 59

The Value of Puppy Raisers' Assessments of Potential Guide Dogs' Behavioral Tendencies and Ability to Graduate
Lara S. Batt, Marjolyn S. Batt, John A. Baguley and Paul D. McGreevy — 71

The Effects of Human Age, Group Composition, and Behavior on the Likelihood of Being Injured by Attacking Pumas
Richard G. Coss, E. Lee Fitzhugh, Sabine Schmid-Holmes, Marc W. Kenyon and Kathy Etling — 77

Volume 22, Number 2

A Review of the Relationship between Indigenous Australians, Dingoes (*Canis dingo*) and Domestic Dogs (Canis familiaris)
Bradley P. Smith and Carla A. Litchfield — 111

Pet, Pest, Profit: Isolating Differences in Attitudes towards the Treatment of Animals
Nicola Taylor and Tania D. Signal — 129

Animal-Assisted Therapy in the Treatment of Substance Dependence
Martin C. Wesley, Neresa B. Minatrea and Joshua C. Watson — 137

Dog-Assisted Therapy in the Treatment of Chronic Schizophrenia Inpatients
Victòria Villalta-Gil, Mercedes Roca, Nieves Gonzalez, Eva Domènec, Cuca, Ana Escanilla, M. Rosa Asensio, M. Elisa Esteban, Susana Ochoa, Josep Maria Haro and Schi-Can group — 149

Preference for, and Responsiveness to, People, Dogs and Objects in Children with Autism
Anke Prothmann, Christine Ettrich and Sascha Prothmann   *161*

Dogs Look Like Their Owners: Replications with Racially Homogenous Owner Portraits
Sadahiko Nakajima, Mariko Yamamoto and Natsumi Yoshimoto   *173*

Self-Reported Comprehension Ratings of Dog Behavior by Puppy Owners
Keven J. Kerswell, Pauleen Bennett, Kym L. Butler and Paul H. Hemsworth   *183*

Volume 22, Number 3

Automation Systems for Farm Animals: Potential Impacts on the Human–Animal Relationship and on Animal Welfare
Cécile Cornou   *213*

An Archaeological and Historical Review of the Relationships between Felids and People
Eric Faure and Andrew C. Kitchener   *221*

"A Gentle Work Horse Would Come in Right Handy": Animals in Ozark Agroecology
Brian C. Campbell   *239*

Relations Among Need for Power, Affect and Attitudes Toward Animal Cruelty
Judith C. Oleson and Bill C. Henry   *255*

Preschoolers' Adherence to Instructions as a Function of Presence of a Dog and Motor Skills Task
Nancy R. Gee, Timothy R. Sherlock, Emily A. Bennett and Shelly L. Harris   *267*

Dolphin-Assisted Therapy: Changes in Interaction and Communication between Children with Severe Disabilities and Their Caregivers
Erwin Breitenbach, Eva Stumpf, Lorenzo v. Fersen and Harald Ebert   *277*

Volume 22, Number 4

The Magic of Animals: English Witch Trials in the Perspective of Folklore
Boria Sax   *317*

The Emergence of "Pets as Family" and the Socio-Historical Development of Pet Funerals in Japan
Elmer Veldkamp   *333*

Reasons for Relinquishment and Return of Domestic Cats (*Felis Silvestris Catus*) to Rescue Shelters in the UK
Rachel A. Casey, Sylvia Vandenbussche, John W.S. Bradshaw and Margaret A. Roberts   *347*

Young Adults' Attachment to Pet Dogs: Findings from Open-Ended Methods
Lawrence A. Kurdek   *359*

Empathic Differences in Adults as a Function of Childhood and Adult Pet Ownership and Pet Type
Beth Daly and L. L. Morton   *371*

Social and Individual Components of Animal Contact in Preschool Children
Manuela Wedl and Kurt Kotrschal   *383*

BOOK NOTICES

A History of Attitudes and Behaviours toward Animals in Eighteenth- and Nineteenth-Century Britain: Anthropocentrism and the Emergence of Animals	292
Animal Capital: Rendering Life in Biopolitical Times	398
Animal Encounters	398
Animals and the Moral Community: Mental Life, Moral Status, and Kinship	90
Between the Species: Readings in Human–Animal Relations	195
Breeds of Empire: The "Invention" of the Horse in Southeast Asia and Southern Africa 1500–1950	291
Ethics and Animal Use	90
Kinship and Killing: The Animal in World Religions	397
Kissing Cousins: A New Kinship Bestiary	91
Noble Brutes: How Eastern Horses Transformed English Culture	398
Social Creatures: A Human and Animal Studies Reader	90
Stalking the Subject: Modernism and the Animal	291
The Death of the Animal: A Dialogue	292
The Dog by the Cradle, the Serpent Beneath: Some Paradoxes of Human–Animal Relationships	292
What Species of Creatures: Animal Relations from the New World	397

BOOK REVIEWS

A Cultural History of Animals	295
Dog: Pure Awareness	308
Knowing Animals	306
Linking Animal Cruelty and Family Violence	95
Swan	401
The International Handbook of Animal Abuse and Cruelty: Theory, Research and Application	403
Victorian Animal Dreams: Representations of Animals in Victorian Literature and Culture	98

CONFERENCES

Animal Behavior Society Annual Meeting	92, 196
Animals: Past, Present and Future	91
ESRC Congress: Re-Inventing the Rural	198
Experiential Learning in Humane Education	91
Human–Animal Interaction: Impacting Multiple Species	93, 198, 292
IAHAIO Conference 2010	199, 293, 399
International Companion Animal Welfare Conference	292
International Conference on Animal-Assisted Therapy	196
International Council for Archaeozoology Conference	199, 293, 399
International Society for Anthrozoology	93, 198, 292, 398
International Society for Applied Ethology	92, 197
International Veterinary Behaviour Meeting	198, 293
Meet Animal Meat	196
Minding Animals	92, 197
Society for the Study of Social Problems	198
UFAW International Symposium	92, 196
Dedication—Bill Balaban	109

Index

NEWS

Call for papers—Special issue of the journal *JAC*	*195*
New Journal—*Humanimalia*	*89*
New Journal—*Mensch und Pferd (Human and Horse)*	*195*
New Web Resource—The Canine in Conversation	*89*
Special Journal Issue on Human–Animal Interaction	*397*

SUBJECT INDEX

Animal Abuse
    factors affecting attitudes to, 255

Animal-Assisted Therapy/Activities
    comparison of live dog with robot-dog, 43
    effects of dolphins on children with severe disabilities, 277
    in the treatment of chronic schizophrenia, 149
    in the treatment of substance dependence, 137

Animal Welfare
    attitudes to animal cruelty, 255
    attitudes to the treatment of animals, 129
    automation systems and farm animals, 213
    reasons for relinquishing cats, 347

Archaeology
    human–cat relationships, 221

Assistance Animals
    puppy raisers' assessments of, 71

Attachment
    to pet dogs, 359

Attacks
    to humans by pumas, 77

Attitudes
    to bats, 19
    to diary cattle, 59
    to farm animals and farming practices, 239
    toward cruelty, 255
    toward the treatment of animals, 129

Bats
    attitudes to, 19

Behavior
    dog; puppy owners' comprehension of, 183
    of children in the presence of a dog, 267
    of people with dementia with real and robot-dogs, 43
    of preschool children in presence of rabbits, 383

Birds
    cognition, 5

Cats
    effect on empathy, 371
    history of human relationships with, 221
    reasons for relinquishment, 347

Cattle
    attitudes to, 59
    effect of stock managers on milk production, 59

Children
    adherence to instructions in the presence of a dog, 267
    factors influencing interactions with rabbits, 383
    with autism; interactions with people, dogs and objects, 161
    with severe disabilities, effects of dolphins on, 277

Companion Animals
    adherence of children to instructions while in the presence of a dog, 267
    and Indigenous Australians, 111
    attachment to dogs, 359
    attitudes to treatment of, 129
    effect on empathy, 371
    effects on oxytocin levels in people, 31
    effects on people with chronic schizophrenia, 149
    effects on people with dementia, 43
    effects on people with substance dependence, 137
    factors influencing preschool children's contact with, 383
    funerals and cemeteries for in Japan, 333
    history of human relationships with cats, 221

interactions with children with autism, 161
looking like their owners, 173
puppy owners' comprehension of dog behavior, 183
reasons for relinquishment of cats, 347

Cruelty, see *Animal Abuse*

Cultural Studies
English witch trials and animals, 317
Indigenous Australians and dingoes and dogs, 111
Ozark farmers' interactions with, and attitudes to, animals, 239
pet funerals and cemeteries in Japan, 333

Dingoes
relationships with Indigenous Australians, 111

Dogs
assisting treatment of chronic schizophrenia, 149
assisting treatment of substance dependence, 137
attachment to, 359
effect on children's adherence to instructions, 267
effect on empathy, 321
effects on oxytocin levels in men and women, 31
interactions with children with autism, 161
look like their owners, 173
puppy owners' comprehension of dog behavior, 183
puppy raisers' assessments of Guide Dogs, 71
relationship with Indigenous Australians, 111

Dolphins
effects on children with severe disabilities, 277

Education
preschool children's interactions with rabbits, 383

Empathy
effect of childhood and adult pet ownership on, 371

Ethics
automation systems and farm animals, 213

Ethology
cognitive, avian, 5

Farm Animals
effect stock managers on, 59
impact of automation systems on, 213
perceptions of by Ozark farmers, 239

Folklore
animals and the English witch trials, 317

Grief, see *Pet Loss*

Health
effect of AAT in the treatment of chronic schizophrenia, 149
effect of AAT in the treatment of substance dependence, 137
effects of dogs on oxytocin levels, 31

History
development of pet funerals and cemeteries in Japan, 333
English witch trials and animals, 317
relationships between Indigenous Australians and dingoes and dogs, 111
relationships between people and cats, 221

Literature
avian consciousness, 5

Older Persons
effects of real and robot-dogs on, 43

Personality
of children, influence on interactions with rabbits, 383

Pet Loss
funerals and cemeteries in Japan, 333

Pets, see *Companion Animals*

Pumas
factors affecting attacks to humans, 77

Rabbits
factors influencing preschool children's contact with, 383

Representations
animals in the English witch trials, 317
avian consciousness, 5

Service Animals, see *Assistance Animals*

Wildlife
dingoes and Indigenous Australians, 111
factors affecting attacks by pumas, 77

AUTHOR INDEX

Articles

Alvarez, S. A., 149
Asensio, M. R., 149
Baguley, J. A., 71
Baño, V., 149
Batt, L. S., 71
Batt, M. S., 71
Bennett, E. A., 267
Bennett, P., 183
Bernstein, P. L., 43
Bertenshaw, C., 59
Bradshaw, J. W. S., 347
Breitenbach, E., 277
Butler, K. L., 183
Campbell, B. C., 239
Casey, R. A., 347
Cornou, C., 213
Coss, R. G., 77
Cuca, 149
Daly, B., 371
DeVoe, D., 31
Domènec, E., 149
Domènech, M., 149
Ebert, H., 277
Escanilla, A., 149
Escuté, M., 149
Esteban, M. E., 149
Etling, K., 77
Ettrich, C., 161
Fančovičová, J., 19
Faure, E., 221
Fitzhugh, E. L., 77
Friedmann, E., 43
Gee, N. R., 267
González, N., 149
Haro, J. M., 149
Hemsworth, P., 183
Henry, B. C., 255
Hickey, M., 31
Karnicky, J., 5
Kennedy, C., 31
Kenyon, M. W., 77
Kerswell, K. J., 183
Kogan, L., 31
Kotrschal, K., 383
Kramer, S. C., 43

Kubialko, M., 19
Kurdek, L. A., 359
Litchfield, C. A., 111
McGreevy, P. D., 71
Medina, J., 149
Miller, S. C., 31
Minatrea, N. B., 137
Morton, L. L., 371
Nakajima, S., 173
Nelson, T., 31
Ochoa, S., 149
Oleson, J. C., 255
Prokop, P., 19
Prothmann, A., 161
Prothmann, S., 161
Ristol, F., 149
Roberts, M. A., 347
Roca, M., 149
Rowlinson, P., 59
Sax, B., 317
Schmid-Holmes, S., 77
Sherlock, T. R., 267
Signal, T. D., 129
Smith, B. P., 111
Soto, V., 149
Stumpf, E., 277
Taylor, N., 129
Vandenbussche, S., 347
Veldkamp, E., 333
Villalta-Gil, V., 149
Von Fersen, L., 277
Watson, J. C., 137
Wedl, M., 383
Wesley, M. C., 137
Yamamoto, M., 173
Yoshimoto, N., 173

Book Reviews

Anderson, P. K., 401
Brandt, K. J., 306
Clutton-Brock, J., 201
Henry, B. C., 95
Irvine, L., 308
Knight, S., 203
McHugh, S., 98
Mitchell, R. W., 295
Reynolds, A., 403

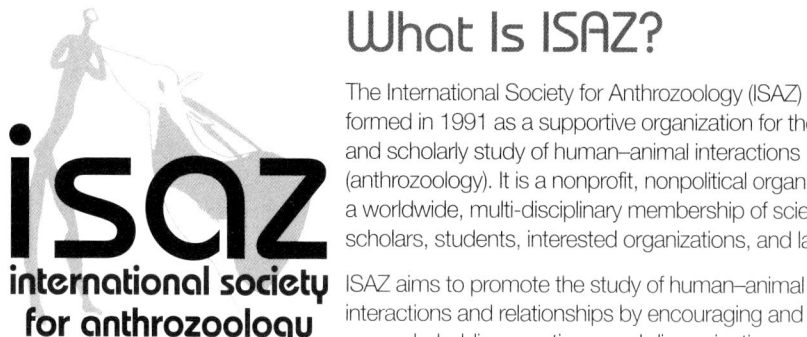

What Is ISAZ?

The International Society for Anthrozoology (ISAZ) was formed in 1991 as a supportive organization for the scientific and scholarly study of human–animal interactions (anthrozoology). It is a nonprofit, nonpolitical organization with a worldwide, multi-disciplinary membership of scientists, scholars, students, interested organizations, and laypersons.

ISAZ aims to promote the study of human–animal interactions and relationships by encouraging and publishing research, holding meetings, and disseminating and exchanging information. To accommodate its international membership, ISAZ has held meetings and conferences across the globe.

Benefits to ISAZ members include:
- The Society's quarterly journal, Anthrozoös, the leading academic journal on human–animal interactions and relationships
- Substantially reduced registration fees for all ISAZ conferences
- Listserv distribution of items of interest to society members
- 20% discount on books and journal subscriptions from Berg Publishers

Anyone can join ISAZ! To ensure that anyone with an interest in the field can enjoy access to the most up-to-date information and scholarship from the field, the Society offers a range of membership options: Individual, Student, Affiliate, and Corporate.

For complete details on membership, visit the ISAZ website today. **www.isaz.net**

Practical guidelines for "One Health" collaborations!

With an innovative approach to the links between animal and human health, **Human-Animal Medicine** addresses the critical health issues facing our world today. With 75% of the disease outbreaks in the last two decades being zoonotic, it is becoming increasingly important for physicians to collaborate with veterinarians and public health professionals to deal with diseases that affect their patients and their patient's pets. This timely guide contains:

- **Concise clinical tips for prevention and treatment** of the H1N1 virus, zoonoses, animal allergy, bites and stings, psychosocial issues, and toxic exposures shared by humans and animals.
- **Protocols and sample forms for professional collaboration** between human health clinicians, veterinarians, and public health professionals.
- **Occupational health guidelines for preventive care of animal workers** including veterinary personnel, farmers, pet store employees, and zoo workers.

Human-Animal Medicine:
Clinical Approaches to
Zoonoses, Toxicants, and
Other Shared Health Risks
*By Peter M. Rabinowitz, MD,
MPH and Lisa A. Conti, DVM,
MPH, DACVPM, CEHP*
December 2009 • 528 pg.
ISBN: 978-1-4160-6837-2

Order today to ensure the most current treatment options for your patients!
Order securely at **elsevierhealth.com** • Call toll-free **1-800-545-2522** • Visit your **local bookstore**

SL90875TR/KB

Journal of the International Society for Anthrozoology

Essential reading for everyone concerned with the broad field of human-animal relations

anthrozoös

A multidisciplinary journal of the interactions of people and animals

Editor-in-Chief
Anthony L. Podberscek, University of Cambridge, UK

Associate Editors
Penny Bernstein, Kent State University, USA
Patricia K. Anderson, Western Illinois University, USA

anthrozoös is a peer-reviewed multidisciplinary journal that provides a vital forum for academic dialogue on human-animal relations. As a pioneer in the field it addresses the characteristics and consequences of interactions and relationships between people and non-human animals across anthropology, ethology, medicine, psychology, veterinary medicine and zoology.

anthrozoös is the official publication of the *International Society for Anthrozoology (ISAZ)*. Join *ISAZ* and benefit from a wide range of membership offers!

For complete details on membership, visit the ISAZ website www.isaz.net.

***anthrozoös* digital archive released**
Access to high quality research covering volume 1 to volume 19 – over 300 invaluable articles previously out of print and unavailable – online at IngentaConnect.

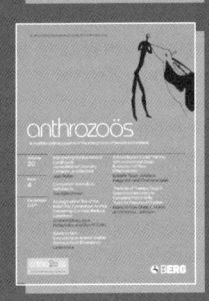

Expand your knowledge, widen your research!
Sign up to RSS feeds and new issue alerts at IngentaConnect.com!

Institutional free trials available at
www.ingentaconnect.com/content/berg

Print ISSN: 0892-7936
E-ISSN: 1753-0377

BergJournals.com

NOTES FOR CONTRIBUTORS

The following details should be used **as a guide only**. Full details can be found at the Berg Publishers web site: www.bergpublishers.com.

Content
Anthrozoös will accept new contributions that describe the characteristics and consequences of interactions/relationships between people and non-human animals. Papers are welcome from the arts and humanities, behavioral and biological sciences, social sciences and the health sciences.

Types of Articles

Commentaries
This section provides a forum for raising issues relating to the fields of interest of the Journal, including theory, methodology, ethics, statistical analysis and nomenclature. Authors may make general points or provide critiques of particular published papers. These articles should usually be no longer than 5000 words.

Review Articles and Research Reports
Reviews—These should address fundamental issues relating to the interactions between people and other animals, and provide new insights into the subject(s) they cover. Original interdisciplinary syntheses are especially welcome. Reviews should be no longer than 8000 words.

Research Reports—both quantitative and qualitative reports are encouraged. These should cover subjects falling within the scope of the Journal and can be up to 6000 words in length.

Note: Word counts do not include tables, figures and references.

Manuscripts
Please submit an original and two other hard copies by post only (see mailing address below; electronic submission is **not** encouraged), along with a cover letter indicating clearly in which section you would like your manuscript to be considered. The two copies must **not** contain authors' names and addresses. After a manuscript has been accepted for publication, the author(s) will be requested to supply an electronic copy of it.

All text (including abstract, notes and references) must be typed (using, preferably, MS-Word), double-spaced, aligned left and printed on one side of each page only. Use active voice whenever feasible, and write in the first person. Tables and Figures should be in Helvetica or Arial font.

Use **American spelling and grammar conventions** throughout, except in non-American quotations and references.

Manuscripts should have line numbers and page numbers throughout. Authors whose first language is not English should have their paper checked by a native English speaker before submitting it.

Manuscript Organization
The title page of the original manuscript should contain the title of the article, and authors' names, affiliations, and present addresses. At the bottom of the page, give the full postal address, phone and fax numbers, and e-mail address of the corresponding author. In the following pages, provide an abstract (250 to 300 words), 3 to 5 keywords (in alphabetical order below the abstract), and the text, including, as appropriate, an introduction, methods, results, discussion, acknowledgements, notes, references, appendices, tables, figure captions, and figures. Each table/ figure must appear on a separate page. The authors' names should appear on the title page of the original manuscript only. Acknowledgements should be left blank until after the paper has been accepted.

Footnotes
Footnotes appear as "Notes" at the end of articles. Authors are advised to include footnote material in the text whenever possible. Notes are to be numbered consecutively throughout the paper and are to be typed, double-spaced at the end of the text (do not use any footnoting or end-noting programs that your text software may offer, as this text becomes irretrievably lost at the typesetting stage).

References
For references in the text, give full surnames for papers by one, two or three authors, but only the surname of the first author, followed by "et al." for four or more (note that "et al." is neither underlined nor italicized). Check that all references in the text are in the reference list and vice versa, and that their dates and spelling match. Check foreign language references particularly carefully for accuracy of diacritical marks such as accents and umlauts.

Cite references in the text as, for example, Swabe (1998) or, if in parentheses, as (Daly and Morton 2006) or (McGreevy, Righetti and Thomson 2005). **Do not** use a comma to separate the author's name from the date. Where more than one paper by the same author has appeared in one year, the reference should be distinguished by "a," "b," "c," etc. (e.g., 1971a). If referring to a specific page in a book, please provide the page number in the citation: for example, (Serpell 1999, p. 45). Where multiple citations are referred to, place in chronological order, from oldest paper to most recent, using a semicolon to separate each reference: for example, (Harrison 1998; Gibbs 1999; Bekoff 2006).

The list of references should be arranged alphabetically by authors' names and chronologically per author. References cited with "et al." in the text should include all authors' names in the reference list. Journal titles should be given in full. References to books or monographs should include editors, edition and volume number, publisher, city and state or country where published, and relevant page numbers. A paper in press may be referenced if it has been accepted for publication. References to personal communications and unpublished work should appear in the text only.

Sample references (note: do not indent):

Galvin, S. and Herzog, H. 1992. Ethical ideology, animal activism and attitudes towards the treatment of animals. Ethics and Behavior 2: 141–149.

Lennon, R. and Eisenberg, N. 1987. Gender and age differences in empathy and sympathy. In Empathy and its Development, 195–217, ed. N. Eisenberg and J. Strayer. New York: Cambridge University Press.

Philo, C. and Wilbert, C. eds. 2000. Animal Spaces, Beastly Places: New Geographies of Human–Animal Relations. London: Routledge.

Paul, E. S. 1992. Pets in childhood: individual variation in childhood pet ownership. Ph.D. thesis, University of Cambridge, UK.

Tables
Each table must be presented on a separate page and be identified by a short, descriptive title placed at the top. Any necessary further explanations (e.g., the results of statistical tests) may be added as footnotes at the base of the table. Make sure that each abbreviation used in a table is fully explained in a footnote. Marginal notations on manuscripts should indicate approximately where tables are to appear. Please use Helvetica or Arial font for all tables. Each table must be cited in the text.

Authors using MS-Word or other word-processing programs must use those programs' table editors to create tables. **Do not** create tables by typing single lines of text followed by a hard return, with spaces or tabs used to align columns.

Figures
All illustrative material (drawings, maps, diagrams, graphs and photographs) should be designated "Figures" and must be cited in the text. For the review process, it is acceptable to supply photocopies of figures. Once a paper is accepted, the author will be required to supply high-resolution files/prints of figures (electronic files are preferred). Figures must be submitted as separate image files and NOT embedded in the word document. Figures will be reproduced exactly as provided. However, as they will be reduced in size to fit the Journal's page format, figures must be of a size which allow a reduction of 50%.

Criteria for Evaluation
Anthrozoös is refereed and papers will be accepted only after appropriate blind review. The general criteria for acceptance are that the research meet standards for publication in a specialty journal appropriate to its field and that it provide new information, sound hypotheses, or insightful analyses relevant to the content area of Anthrozoös. This is a multidisciplinary journal, and authors should be aware that their own discipline's jargon may be unfamiliar to readers from other disciplines. Please keep jargon to a minimum and provide a complete methods section. If you are in doubt about this, please err on the side of providing fuller explanations. The Editor can always cut material but cannot add it.

Ethical Considerations
Studies Involving Animals
If studies have the potential to compromise animal welfare, precautions should be taken to reduce possible harm to the animals involved. Authors should identify welfare concerns and describe the measures that were taken to mitigate animal pain or distress. Anthrozoös will not accept any manuscripts based on research inflicting suffering or cruelty on animals.

Studies Involving Humans
Informed consent should be given by persons participating in the studies reported. Any sensitive data should be handled with confidentiality and stored securely. When reporting results, participants should remain anonymous.

Conflicts of Interest
Any personal, financial, or other potential or actual conflicts of interest relating to the study should be conveyed by the authors.

Copyright
Papers are accepted on the understanding that they are subject to editorial revision and that they are contributed only to this Journal. Copyright in the article, including the right to reproduce the article in all forms and media, shall be assigned exclusively to the Journal. The transfer of copyright to Anthrozoös takes effect when the manuscript is accepted for publication.

Proofs
One set of proofs will be sent to the corresponding author as an e-mail attachment (PDF). Only typographic errors may be corrected at this stage.

On publication, authors will be sent a PDF e-print (with nonprinting watermark) of the final, published version of their article for personal use, and will be able to order a free copy of the issue in which their article appears through Berg Publishers. Contact Ken Bruce at kbruce@bergpublishers.com.

Mailing Address
All manuscripts should be sent to:
Dr. Anthony Podberscek, Editor Anthrozoös
University of Cambridge
Department of Veterinary Medicine
Madingley Road
Cambridge CB3 OES, UK

Manuscripts will be acknowledged and entered into the review process (described above).

Queries about any of the guidelines can be sent to the editor via e-mail: alp18@cam.ac.uk.